CIVIL ENGINEERING SPECIFICATION FOR THE WATER INDUSTRY

(SEVENTH EDITION)

MARCH 2011

Published by WRc plc
Frankland Road
Blagrove
Swindon
Wiltshire
SN5 8YF

on behalf of UK Water Industry Research Ltd
1 Queen Anne's Gate
London
SW1H 9BT

Copyright

ISBN 978 1898920 687

Feedback Arrangements

Technical queries should be referred to: CESWI Technical Secretary, WRc plc, Frankland Road, Blagrove, Swindon, Wiltshire, SN5 8YF

Tel: + 44 (0) 1793 865105
Fax: + 44 (0) 1793 865001
Web: http://ceswi.wrcplc.co.uk

Further Copies

Further copies can be obtained from: WRc plc, Frankland Road, Blagrove, Swindon, Wiltshire, SN5 8YF

Tel: + 44 (0) 1793 865012
Fax: + 44 (0) 1793 514562
Email: publications@wrcplc.co.uk
Web: www.webookshop.com

FOREWORD - SEVENTH EDITION

The Civil Engineering Specification for the Water Industry (CESWI) was first published in July 1978, followed by second, third, fourth, fifth and sixth editions in August 1984, May 1989, October 1993, July 1998 and June 2004, respectively. It has become the standard document for civil engineering contracts let by the Water Undertakers and Sewerage Undertakers in England, Wales and Scotland.

In 2010, a national survey was undertaken to elicit comments on how the sixth edition of the Specification might be improved. Announcements were subsequently made in appropriate publications and comments were invited from the Water Industry.

In redrafting CESWI, the Working Group adhered to the principles used in the earlier editions; in particular, that the document should be performance based and that, in letting any contract for civil engineering works, the several documents that form the Contract are to be taken as being mutually explanatory of one another. CESWI does not, therefore, repeat other nationally accepted Specifications, rather it refers to them in the relevant Notes for Guidance and in the Appendices.

As in the previous edition, the seventh edition is generic in form and not tied to any particular form of Contract or method of valuation.

As there is no consistent terminology across the Conditions of Contract most commonly used in the Water Industry, the term "Client" has been adopted, as a classification independent of the specific conditions of contract, for the terms Employer, Authority and Purchaser. Similarly, wherever possible, the seventh edition has been re-worded to remove specific reference to the "Contractor" or "Contract Administrator", in order to make the seventh edition more flexible with respect to the many alliancing, joint ventures, and other contracting entities that now carry out work for the Water Industry.

There is no reference in the Specification to satisfy legislation. This follows the general principle that the Works are to be designed and constructed to comply with all relevant Statutes. A number of Notes for Guidance have been included, where appropriate.

As with previous editions, it is envisaged that additional Clauses will be added to provide for individual requirements and the inclusion of such Clauses has been allowed for in the system of numbering, which should be followed when adding any additional Clauses in order to maintain consistency of presentation. This will be further facilitated by the availability of an electronic version (see www.wrcplc.co.uk/ceswi).

CESWI 7 is a comprehensive, universal and relevant Specification for the Water Industry. It is compatible with most nationally produced Engineering Forms of Contract. It remains a valuable foundation document for use throughout the UK Water Industry.

The Specification was reviewed and redrafted by the following Working Group:

Mike Smith (Chairman)	Bristol Water
Santoro Salvatore	Anglian Water
Les Gonzalez	Dŵr Cymru
Dominic Wall	Dŵr Cymru
Mick Abbott	Northumbrian Water
Iain MacDonald	Scottish Water
Mike Fairlamb	Severn Trent Water
John Foster	Severn Trent Water
Kate Harrison	South East Water
Brian Spencer	Southern Water
Jack McCarey	South West Water
Don Ridgers	Thames Water
Duncan McKinley-Bourne	United Utilities
Brian Spark	Water UK
Stuart Lewis	Wessex Water
Brian Smith	Yorkshire Water
Sarah Wilcox (Technical Secretary)	WRc plc
Jo Hulance	WRc plc
Nick Orman	WRc plc
Julia Trew	WRc plc
Huw Williams	WRc plc

Cover designed by WRc plc.

CONTENTS

INTRODUCTORY NOTES

1. The Foreword, Product Specifications and Levels of Attestation, Notes for Guidance and Appendices are not part of the Specification and do not affect the interpretation either of the Specification or of any other Contract documents.

2. The following principles, which reflect current engineering practice within the Water Industry, have been followed in drafting this substantially revised edition of the Specification:

   (i) The Specification is written primarily in terms of the performance required, leaving the Constructor, so far as possible, free to decide his method of working.

   (ii) The document is intended for use with most engineering types of Contract within the Water Industry for the construction of civil engineering works. As such, it is not associated with any particular Conditions of Contract.

   (iii) The Specification is generic in nature and is intended for use in conjunction with whatever method of valuation is used.

   (iv) There is no reference in the Specification either to the method of assessing payment or to who should bear specific costs. These matters are left for inclusion in the relevant Contract.

   (v) Wherever other explicit and relevant documents are available, they are referred to rather than repeated in the Specification.

   (vi) References to Acts of UK Parliament and Statutory Instruments are generally omitted from the main text (though not from the Notes for Guidance), since compliance with Statute is a general legal requirement.

   (vii) Traditional requirements which are desirable but impossible to achieve in a literal sense have been omitted (e.g., “pipes to be laid true to line and level”).

   (viii) All actions are to be carried out by the Constructor unless otherwise stated in the Specification.

3. Notes for Guidance are printed in the margins beside the text to which they refer and are designed primarily to assist those preparing Contract documents. One of their main purposes is to explain apparent omissions from the text, such as those referred to in Note 2 above, and to cross-refer to other documents where such matters are covered.

4. Acts of Parliament, Statutory Instruments and other documents quoted in the Notes for Guidance and the Appendices are those in force at the time of publication.

5. Reference to Clause Numbers in the main text and in the Notes for Guidance relate to Clause Numbers of this Specification.

6. It is recommended that the boundaries and limitations of the Works included in the Contract are described at the beginning of the Specification for the particular Contract, and should include items such as:

   (i) The Site and access; drawing reference showing boundaries.

   (ii) Sections of the Works; full description.

   (iii) Other Contracts; lists of work content, areas and timings.

   (iv) Operational restrictions; Site rules and requirements.

   (v) Client-supplied materials; list of what, when and from whom.

   (vi) Constructor design; definition of the parts that the Constructor is to design and the standard(s) required.

7. All Works should be carried out with the objective of causing a minimum amount of disturbance to the environment, especially where there is a large environmental interface.

PRODUCT SPECIFICATIONS AND LEVELS OF ATTESTATION

The Utilities Directive (2004/17/EC) on the Procurement Procedures of Entities Operating in the Water, Energy, Transport and Postal Services Sectors came into force on 30 March 2004. Article 34 of this Directive details the requirements of this Directive regarding the use of Standards.

When specifying the requirements of any product, the following procedure is used:

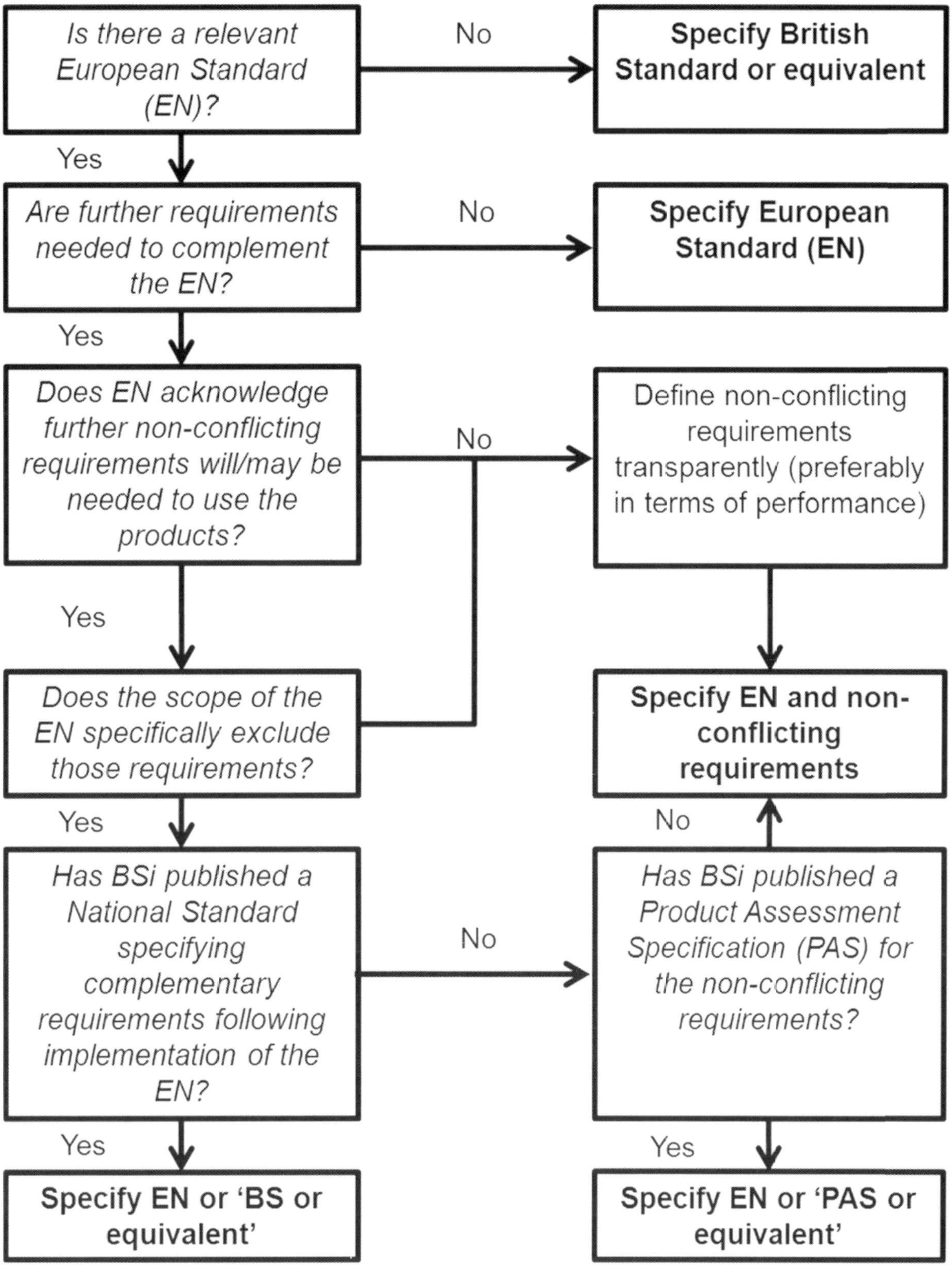

A relevant EN is any EN covering the subject in question which is in force in the EU. In the UK, ENs are published as BS ENs. A Water Industry Specification may be used where there is no relevant EN, BS or equivalent available.

A European Standard, British Standard or equivalent will not normally be available in the case of newly-developed products but this is not to be seen as inhibiting the use of such products. Water Industry Specifications, issued by the Water UK Standards Board, deal with products such as these and are called up in the Specification. However, in the case of any innovatory product for which no European Standard, British Standard or Water Industry Specification (or the equivalent of either) exists, care should be taken to ensure that the product is fit for purpose. This might be achieved by using products which have been assessed by an independent body. The British Board of Agrément (BBA) is authorised to issue European Technical Approvals under the provisions of the Construction Products Directive (89/106/EEC).

Under the provisions of the Construction Products Directive (89/106/EEC), the European Commission decides the appropriate level of attestation when a harmonised European Specification is published under a mandate given in connection with that Directive. The Specification, therefore, now provides for Levels of Attestation less stringent than full Third Party Certification, where so decided by the Commission.

Where the product is also covered by a Mandate under the provisions of the Construction Products Directive (89/106/EEC), the relevant system of conformity attestation chosen by the Commission of the European Community (CEC), and leading to the use of the CE mark for a product, applies to those elements of the product specification which relate to the Mandate. Third Party Certification is recommended as a cost-effective means of ensuring compliance with Standards. However, Water Industry policy is for individual purchasers to determine where Third Party Certification is required.

Additional quality assurance requirements, including Third Party Certification, may be sought by the purchaser as a cost-effective means of ensuring compliance with Standards. BSi Kitemarking is an example of Third Party Certification.

Currently, where European Standards refer to requirements for the effect of materials on potable water quality, National Regulations apply. Before accepting any material which will come into contact with potable water or water to be used for potable supply, the Client will have regard to the provisions of:

a) For England:

Water Supply (Water Quality) Regulations 2010 (SI 2010 No. 994 (W.99)).

(b) For Scotland:

Water Supply (Water Quality) (Scotland) Regulations 2001 (SI 2001/207).

(c) For Northern Ireland:

Water Supply (Water Quality) Regulations (Northern Ireland) 2002 (SR 2002/331).

Reference should be made to the current list of approved substances prepared by the Drinking Water Inspectorate.

Reference should also be made to the 'Water Fittings and Materials Directory', published by the Water Regulations Advisory Scheme Ltd on its website (www.wras.co.uk/dorectory).

Requirements for the testing of materials for use in contact with potable water are dealt with in:

i) DD 256 for metallic materials.

ii) BS 6920-1 for non-metallic materials.

Due consideration should be given to selecting materials which promote sustainability and minimise environmental impact.

SECTION 1

GENERAL

The Notes for Guidance are not part of the Specification

1.1 DEFINITIONS

(i) The Contract could be a construction agreement e.g., NEC or IChemE, or an adoption agreement, e.g., Self Laying of Water Mains and Services.

1. This Specification is written on the assumption that the Party entering into a Contract that includes this Specification will be bound by the Specification terms and be responsible for following its provisions.

2. "Party" means a company, contractor, alliance, joint venture, consortium, etc.

3. "Client" means the person or persons, firm, company or other body for whom the Works are to be constructed and includes the Client's personal representatives, successors and permitted assignees.

4. "Environmental Regulator" means in England and Wales, the Environment Agency (EA), in Scotland, the Scottish Environmental Protection Agency (SEPA) and in Northern Ireland, the Northern Ireland Environment Agency (NIEA).

5. "River Authority" means in England and Wales, the Environment Agency (EA), in Scotland, the Scottish Environmental Protection Agency (SEPA) and in Northern Ireland, the Department for Agriculture and Rural Development.

6. The "Highway Reinstatement Specification" means in England, 'Specification for the Reinstatement of Openings in Highways', Department for Transport and published by HAUC. In Scotland it means the 'Specification for the Reinstatement of Openings in Roads' the Scottish Government. In Wales it means 'The Specification for the Reinstatement of Openings in Highways (2nd Edition), Welsh Assembly Government.

1.2 ACCOMMODATION FOR THE CONTRACT

(i) Consent of the Local Planning Authority may be required for Site accommodation.

(ii) BS 6767-1 deals with transportable accommodation units.

(iii) Reinstatement is covered in Clause 3.9.

(iv) Details of accommodation and disposal of all domestic types of waste should be stated in the Contract.

(v) Any special insurance requirements should be a Special Condition of Contract.

(vi) The siting and availability of all accomodation should be stated in the Contract.

1. Until the completion of the Works, accommodation and parking facilities, as described in the Contract, shall be equipped and maintained. Accommodation shall be erected, furnished, equipped, fully serviced and ready for use within 7 days of the work commencing on Site.

2. Where movable offices are required by the Contract, these shall be relocated as directed under the terms of the Contract.

3. Where the Contract requires telephone facilities to be provided, such facilities shall have separate connections with privacy of conversation.

4. All Site accommodation, services and parking shall be removed upon completion of the Works or at such other date as directed under the terms of the Contract.

1.3 BILLPOSTING AND ADVERTISING

(i) Notice boards or scheme signboards required by the Client should be detailed in the Contract.

(ii) Notice boards and signboards should be erected and maintained by the

1. Billposting or advertising of any kind shall not be undertaken, unless otherwise stated in the Contract.

Contractor at a location and in the form, layout and number as specified in accordance with the Contract.

(iii) Consent of the Local Planning Authority may be required for notice boards or signboards.

1.4 BRITISH STANDARDS AND OTHER DOCUMENTS

(i) Any particular requirements should be described in the Contract after taking due account of the guidance in the Product Specifications and Level of Attestation at the front of CESWI.

(ii) Where appropriate, use should be made of any technical and advisory services offered by manufacturers.

1. British Standards and other documents referred to in the Contract shall be deemed to be those current 28 days prior to the date for return of Tenders.

2. Any reference in the Contract to a Standard published by the British Standards Institution, or to the Specification of another body, shall be construed equally as reference to an equivalent one.

1.5 TIDINESS OF SITE

(i) Special storage requirements should be described in the Contract.

1. Materials, plant and equipment shall be positioned, stored and stacked in an orderly manner.

2. All pipes and fittings shall be stored off the ground, in a clean environment to prevent any contamination of materials prior to their use.

3. All Site waste and surplus shall be removed from Site.

4. Pipe end caps shall be fitted at either end of the pipe, until the pipe is used in the Works, to prevent vermin entering.

1.6 ENTRY ONTO THE SITE

(i) The Contract should prescribe the extent of the Site, including working areas, accesses and the periods for which they will be available. Reference should be made to any known hazards.

(ii) The Client may have to obtain planning permission where an access from a classified road or street is specified.

(iii) It should be assumed that formal entry notices have already been served by the Client, and owners and occupiers have been alerted to impending entry.

(iv) The Contract should describe any Code of Practice or other relevant documentation in connection with pipelaying on private land.

(v) Any requirements for entry on third party land, including any notice required, should be specified in the Contract.

1. The Client shall, at all reasonable times, have access to the Works and to the Site and to all workshops and places where work is being prepared or from where materials, manufactured articles and machinery are obtained for the Works, and every facility for, and every assistance in, obtaining such access or the right to such access shall be afforded.

2. Records shall be kept of the dates of the entry onto and departure from all property and lands of each owner and occupier.

3. Records shall be kept of the dates of the erection and removal of all enclosures.

4. Copies of these records shall be provided as specified in the Contract. Copies of similar records in respect of roads, footpaths and thoroughfares shall be kept.

5. No part of the Site shall either be entered or used for any purpose unconnected with the Works.

1.7 SURVEY OF HIGHWAYS, PROPERTIES AND LAND

(i) The composition and extent of the survey should be detailed in the Contract and may include photographs, video recordings, sketches and professional reports.

1. Prior to entry, and where appropriate, surveys shall be carried out of the condition of highways, properties, lands including trees, boundaries, crops and any other features which may be affected by the Works.

2. The survey shall be carried out with the responsible Highway Authority and owners or occupiers.

(ii) The Contract should describe any special requirements for land used for organic farming.

(iii) The format of the survey record photographs should be stated in the Contract.

3. A representative selection of photographs shall be provided in a digital format (CD or DVD) for the purpose of the survey record. Each photograph shall be uniquely numbered and dated.

4. All photographs/records shall be signed by the landowner as a true record of the condition of the property and land, prior to the commencement of the Works.

1.8 LEVELS AND REFERENCE POINTS

(i) The Contract should prescribe the datum level for the Works, together with any master benchmarks. Precise reference of the Works to existing features or to the Ordnance Survey National Grid should be shown.

(ii) Local differences can arise between GPS and OS coordinates.

1. Details shall be provided of the level and location of the temporary benchmarks and reference points which are proposed to be used.

1.9 TEMPORARY SITE FENCING AND GATES

(i) The location and type of fencing including access and gates, should be detailed in the Contract.

1. The Site shall be adequately fenced.

2. Where the type and locations of temporary site fencing and gates are described in the Contract, such fencing and gates shall be erected before commencing any other work in that portion of the Site.

3. All Site fencing and gates shall be regularly inspected and maintained, and any defects made good without delay. Access shall be provided in temporary Site fencing and gates, as necessary, for the use of the occupiers of adjacent lands.

4. Temporary Site fencing and gates shall remain in position until either they are replaced with permanent fencing and gates, or the Works are sufficiently completed to enable that portion of the Site to be brought into use.

1.10 INTERFERENCE WITH LAND INTERESTS

(i) All land required for the Works should be prescribed in the Contract. Additional land or interests found necessary to carry out the Works may need to be obtained, including any associated planning consent.

(ii) The Contract should describe any special precautions necessary to comply with the Client's obligations.

(iii) The availability and location of accommodation and storage areas outside the Site should be stated in the Contract.

1. Constructional operations shall be confined within the Site, or such other areas of land as may be negotiated. No trespass or breach of any rights, bye-laws or regulations shall take place.

2. Subject to any unavoidable disturbance which may be necessitated by the execution of the Contract, any rights which may be enjoyed on or near the Site shall not be interfered with.

3. Before exercising any right negotiated in connection with wayleaves, easements or accommodation outside the Site, written notification of such arrangements shall be provided prior to entry.

1.11 INTERFERENCE WITH ANY ACCESS TO PROPERTY, APPARATUS OR SERVICE

1. The access requirements shall be identified in the Contract and provided in advance of any property, apparatus or service being affected by the Works. Alternative arrangements shall be provided before interfering with any access to property, apparatus or service. Affected owners and occupiers shall be notified, in writing, 14 days in advance of any such interference and it shall be confirmed that alternative arrangements have been agreed.

2. Should it become impractical to maintain vehicular access to any property, apparatus or service at any time during the construction of the Works, alternative arrangements shall be provided and maintained. Every

assistance shall be provided to the owner/occupier or tenant affected by the Works to enable them to undertake all aspects of their normal activity.

3. The access and service requirements of those with special needs shall be taken into account.

4. Work in the highway shall be planned and carried out to minimise the disruption to access for local residents. Safe pedestrian access to all properties shall be maintained at all times.

1.12 PROCEDURE FOR COMPLAINTS AND CLAIMS

(i) The Contract should define the meaning of "promptly" in accordance with the service objectives of the Client.

(ii) The Contract should define any specific requirements for logging and dealing with complaints and correspondence, in line with Regulatory requirements.

1. Immediately following any damage or injury arising from the execution of the Works, written notification shall be provided without delay to the Client.

2. Details of all complaints, claims or warnings of intended claims received from third parties shall be notified without delay to the Client.

3. Any complaints, claims, damage or injury by owners or occupiers shall be dealt with promptly.

1.13 PROTECTION AGAINST DAMAGE

(i) Any permanent support known to be required should be described in the Contract.

(ii) NJUG (National Joint Utilities Group) Publication Volume 4 'Guidelines for the Planning, Installation and Maintenance of Utility Services in Proximity to Trees' gives guidance on working near trees.

1. All necessary precautions shall be taken to avoid causing any unwarranted damage to highways, roads, properties, lands, trees, roots, crops, boundaries and any other features, and the apparatus of Statutory Undertakers, the Highway or Roads Authority and other parties.

2. Where any portion of the Works is close to, across or under any existing apparatus of Statutory Undertakers, the Highway or Roads Authority or other parties, temporary support shall be provided and work carried out around, under or adjacent to all apparatus in a manner designed to avoid damage, leakage or danger, and to ensure uninterrupted operation.

3. The Client and the Statutory Undertaker, Highway or Roads Authority or owner concerned, as appropriate, shall be notified should any leakages or damage to existing services, highways or roads be discovered, and every facility shall be afforded for the repair or replacement of the apparatus affected.

1.14 USE OF HERBICIDES AND PESTICIDES/PROTECTION OF AGRICULTURAL LAND

(i) Where necessary, the Contract should detail the requirements to prevent the spread of agricultural, animal and poultry diseases, including any restrictions on movement of materials.

(ii) The document 'Every Drop Counts – Keeping Water Clean' by the Crop Protection Association gives guidelines on the use of pesticides.

1. Details of all herbicides and pesticides proposed for use shall be submitted to the Client for approval.

2. All pesticides shall be selected from the list of approved pesticides published by DEFRA, the Scottish Government and the Welsh Assembly.

1.15 WORKS AFFECTING WATERCOURSES

(i) The Client should obtain all necessary Statutory consents relating to the permanent Works.

(ii) Any requirement to liaise with the appropriate Land Drainage Authority (or equivalent) should be described in the Contract.

(iii) The following Statutory provisions may also be relevant:

1. Written notification shall be provided 14 days in advance of any intention to start any part of the Works affecting a watercourse, canal, lake, reservoir, borehole, aquifer or catchment area.

2. Watercourses including land and/or road drainage within the Site shall be maintained in effective working condition at all times.

3. All practicable measures shall be taken to prevent the deposition of silt or other material in, and the pollution of, or damage to, any existing watercourse, canal, lake, reservoir, borehole, aquifer or catchment area arising from operations and acts of vandalism.

(a) Impeding Flow in a Watercourse: Water Resources Act 1991, Section 107, and Land Drainage Act 1994,

(b) Pollution of a Watercourse or Underground Strata: Water Resources Act 1991, Sections 85, 86 and 90, Salmon and Fresh Water Fisheries Act 1975, Section 4;

(c) Environmental Protection Act 1990 and Environment Act 1995; and

(d) Wildlife and Countryside Act 1981.

(iv) It should be noted that the requirements of the Environment Agency or SEPA for temporary and/or permanent Works could have implications for the programme.

(v) Where the Site impinges on an Aquifer Protection Zone, this should be stated in the Contract.

(vi) Useful guidance can also be found in the following documents, CIRIA C650 handbook 'Environmental Good Practice on Site' and CIRIA C649 and C648 'Control of Water Pollution from Linear Construction Projects'.

(vii) The Environment Agency, SEPA and NIEA provide advice to industry and the public on their legal responsibilities and good environmental practice through a series of Pollution Prevention Guidance Notes (PPGs) (see www.environment-agency.gov.uk).

4. Unless otherwise specified, approval shall be obtained for all temporary discharges, crossings or diversions to watercourses from the appropiate Environment Regulator or River Authority and the work shall comply in all respects with their requirements.

5. Any construction equipment and vehicles which present a risk of affecting a watercourse shall be removed from Site.

1.16 WATER SUPPLY HYGIENE

(i) The use of the term "pathogenic organism test" has been avoided since other tests will be involved.

(ii) If any particular requirements or local provisions apply to any requirements in connection with Clause 1.16.4, then these should be detailed in the Contract.

(iii) Areas of the Works which are defined Areas of Restricted Operations should be detailed in the Contract.

(iv) Particular water supply hygiene requirements of the Client may be included in the Contract.

1. All Works relating to the installation of water mains and/or services shall be carried out in accordance with the Water Undertaker's Drinking Water Hygiene Code and Water UK's 'Principles of Water Supply Hygiene and Technical Guidance Notes (www.water.org.uk).

2. Any person engaged in work which is defined by Water UK's 'Principles of Water Supply Hygiene: Technical Guidance Notes' as "Restricted Operations" shall be registered by the National Water Hygiene Scheme administered by Energy and Utility Skills (www.eusr.co.uk) and carry a National Water Hygiene Card.

3. Any person engaged in work which is defined by Water UK's 'Principles of Water Supply Hygiene: Technical Guidance Notes' as a "Restricted Area" shall be informed of the need for personal hygiene and risk of contamination, and be issued with the Water Undertaker's Drinking Water Hygiene Code.

4. As required by the Water Undertaker's Drinking Water Hygiene Code, written notification shall be provided to the Water Undertaker of personnel working on "Restricted Operations" who develop symptoms of certain infectious diseases.

1.17 APPARATUS OF STATUTORY UNDERTAKERS, HIGHWAY OR ROADS AUTHORITY AND OTHERS

(i) The Contract may indicate what is believed to be the details of the apparatus of Statutory Undertakers, Highway or Roads Authorities and others close to the Works. If these details are indicated then no warranty is given as to the accuracy or completeness of this information.

(ii) Water UK and British Gas (currently Transco) have agreed a 'Model Consultative Procedure for Pipeline Construction Involving Deep Excavation', published in 1993.

(iii) The information given under Clause 1.17.1 would not normally include connections to and from premises and street furniture.

(iv) Responsibility for carrying out diversion or removal of apparatus should be detailed in the Contract.

(v) Any requirements of Statutory Undertakers or public bodies should be described in the Contract, together with contact names, locations, telephone and fax numbers.

(vi) For Works likely to affect other apparatus in a street or road see the New Roads and Street Works Act 1991, or the Street Works (Northern Ireland) Order 1995.

1. Before designing (where relevant) or commencing any excavations, liaison with all relevant Statutory Undertakers, the Highway or Roads Authority and other owners of apparatus shall take place to ensure satisfaction as to the exact position of existing apparatus which may affect, or be affected by, the construction of the Works.

2. Written notification shall be provided in advance of any diversion or removal of apparatus which may be required for convenience or because of the proposed method of working.

3. A drawing of all services and apparatus encountered shall be provided to the Client. It shall highlight any differences with the information provided by the Statutory Undertaker and Highway or Roads Authority.

4. The HSE Guidance Notes, Statutory Undertakers and private company requirements shall be fully complied with when working in the vicinity of their apparatus.

1.18 TRAFFIC REQUIREMENTS

(i) Chapter 8 of the 'Traffic Signs Manual' covers many aspects which are sometimes found in Specifications, such as traffic signals and signs, one-way working and minimum carriageway widths.

(ii) The Contract should state who is responsible for:

(a) road closures;

(b) road diversions;

(c) New Roads and Street Works Act notices;

(d) temporary traffic control measures - including noticeboards; and

(e) any local New Roads and Street Works Act requirements.

(iii) For general requirements relating to road closures and diversions, see Paragraph 8.53 of Chapter 8 of the 'Traffic Signs Manual'.

1. There shall be full compliance with the Code of Practice 'Safety at Street Works and Road Works' issued by the Secretaries of State for Transport, Scotland and Wales under Sections 65 and 124 of the New Roads and Street Works Act 1991 as amended by the Traffic Management Act 2004, and the Department of Environment (Northern Ireland) under Article 25 of the Street Works (Northern Ireland) Order 1995. Additionally, there shall be full compliance with the relevant provisions of the Traffic Safety Measures for Road Works, as contained in Chapter 8 of the 'Traffic Signs Manual' (http://www.dft.gov.uk/).

2. Before any work in, or affecting the use of, any highway or road is commenced, the proposed method of working, including any special traffic requirements, shall be agreed with, and confirmed in writing to, the Client, the Highway or Roads and Police Authorities.

3. All construction concerning Works in, or access to, the highway or road shall be carried out in co-operation with the Highway or Roads and Police Authorities.

4. Where the diversion or closure of any existing carriageway, footway or public right of way is temporarily necessitated by the Works, then an alternative shall be provided and maintained which shall be operational before any interference with the existing way takes place.

5. Where ramps are required, they shall be provided and maintained to a standard suitable in all respects for the class or classes of traffic or pedestrians requiring to use them.

(iv) Arrangements made with the Highway or Roads and Police Authorities may include provisions for essential users, such as buses and refuse collection.

(v) Consultation and coordination with the Highway Authority may require specific timescales for planning and coordination. Details should be provided in the Contract.

(vi) Direction 34(1)(b) of the 'Traffic Signs Regulations and General Directions 1994 No. 1519' and the 'Specification for Portable Traffic Signal Control Equipment for use at Roadworks TR 2502' should be referred to.

6. All reasonable steps shall be taken to prevent vehicles entering and leaving the Site depositing mud or other debris on the surface of adjacent roads or footways, and any materials so deposited shall be removed expeditiously.

7. An up-to-date list of supervisors and operatives who have achieved accreditation in the relevant activities specified in the New Roads and Street Works Act 1991 as amended by the Traffic Management Act 2004 or the Street Works (Northern Ireland) Order 1995 shall be maintained on Site.

8. Emergency vehicle access to all properties shall be maintained at all times.

9. The Works shall be planned and executed to ensure that all relevant notices are submitted within the required times, and such that relevant coordination activities with the Highway Authority can be carried out.

10. Where restricted working hours apply, all excavations in the section of highway to which the restriction applies shall be backfilled and reinstated to an interim standard or covered with a road plate where the Highway Authority permits.

11. Applications shall be submitted to the appropriate Highway Authority for approval to use portable traffic signal equipment, as necessary, on each occasion.

1.19 EMERGENCY ARRANGEMENTS

(i) The Contract should indicate the responsibility for Site security.

1. Arrangements shall be maintained whereby labour, materials and equipment can quickly be called out, outside normal working hours, to carry out any work needed for an emergency associated with the Works. An up-to-date list of addresses and telephone numbers of the staff who are currently responsible for organising emergency work shall be maintained.

2. Employees shall be made aware of any relevant arrangements, including those of the Client, which are in existence for dealing with emergencies.

1.20 HAZARDOUS SUBSTANCES

(i) The following Statutory and other provisions are also relevant:

- *Storage and Use of Explosives: Construction (General Provisions) Regulations 1961, Part VI, Explosives Acts 1875 and 1923 and Orders in Council Nos. 6, 6A, and 6C, made under the 1875 Act and Control of Explosives Regulations 1991;*
- *Storage of Petroleum: Petroleum Spirit (Motor Vehicles etc.) Regulations 1929, Petroleum (Consolidation) Act 1928 and Petroleum (Consolidation Act (Northern Ireland) 1929;*
- *The Dangerous Substances and Explosive Atmospheres Regulations 2002 and The Dangerous Substances and Explosive Atmospheres Regulations (Northern Ireland) 2003;*
- *The Planning (Hazardous Substances) Regulations 1992 (as amended for England, Scotland and Northern Ireland);*
- *The Planning (Control of Major-Accident Hazards) Regulations 1999, the Planning (Control of Major-Accident Hazards) Regulations (Northern Ireland)*

1. Hazardous substances shall not be brought onto the Site, used for any purposes or incorporated into the Works without the prior written consent of the Site owner, unless specified in the Contract. All necessary licences shall be obtained.

2. The location of each explosives magazine and store of any other hazardous substance on the Site shall be approved in writing.

3. The storage of blasting explosives shall be in accordance with the conditions (if any) of the Statutory licence obtained and the relevant provisions of BS 5607.

2000, the Planning (Control of Major-Accident Hazards) Regulations (Scotland) 2009; and

- *The Control of Pollution (Oil Storage) (England) Regulations 2001 and the Control of Pollution (Oil Storage) (Northern Ireland) Regulations 2010.*

1.21 ENVIRONMENT AND SUSTAINABILITY

(i) Copies of the Client's policy should be included in the Contract.

(ii) The following publications are available from CIRIA: 'Environmental Handbooks for Building and Civil Engineering Projects' and 'Environmental Impact of Materials'.

(iii) The Contract should state any requirements in connection with environmental or ecological surveys required to be undertaken by the Contractor (or any other party) prior to the start of construction.

(iv) Guidance is available in the 'Pollution Prevention Guidelines' (http://publications.environment-agency.gov.uk/).

(v) Details of any known controlled wastes that require removal from Site should be included in the Contract.

(vi) Guidance on recycling opportunities can be obtained from WRAP (www.wrap.org.uk).

(vii) Guidance on materials management plans can be obtained from the CL:AIRE Code of Practice (www.claire.co.uk).

(viii) Work in and around hedgerows should comply with the requirements of the Hedgerows Regulations 1997.

(ix) Specific requirements of planning permission or environmental risk assessments should be specified in the Contract.

(x) All relevant guidance should be followed that is issued by the environmental regulator in connection with pollution prevention.

(xi) Reference should be made to the Site Waste Management Plan Regulations 2008.

1. Design and construction of the Works shall be carried out in accordance with any environmental mitigation measures specified in the Contract.

2. All Site operations shall be managed to minimise waste of construction materials and maximise the recycling of waste.

3. The Contractor shall make his Site staff and contractors fully aware of any specific environmental practices relevant to the Site, including the process for reporting environmental incidents specified in the Contract.

1.22 CUSTOMER CARE

(i) Copies of the Client's policy should be included in the Contract.

1. All construction operations shall be carried out with due regard for the customer care policy specified in the Contract.

1.23 ASSET RECORDS

(i) Details of the Client's requirements should be included in the Contract.

1. As-built records of all Works constructed, and existing services encountered during the construction of the Works, shall be submitted in accordance with the Contract

1.24 TRAINING

(i) Details of the Client's requirements should be included in the Contract.

1. The Client's staff, or appointed representatives, shall be trained in the operational and maintenance requirements for any new installation. This training shall be provided just prior to take over and at times to suit the Client's staff availability.

SECTION 2

MATERIALS

The Notes for Guidance are not part of the Specification.

(i) The use, installation, application or fixing of materials and components should be in accordance with the manufacturer's recommendations.

2.1 MATERIALS IN CONTACT WITH POTABLE WATER

(i) Where possible, all fittings used as part of the supply of treated water for domestic purposes in premises should be approved by WRAS.

(ii) See the 'Principles of Water Supply Hygiene: Technical Guidance Notes' published by Water UK (www.water.org.uk).

(iii) When planning, designing and constructing water mains and/or services in brownfield sites, reference should be made to the UKWIR document "Guidance for the Selection of Water Supply Pipes to be used in Brownfield Sites".

1. All materials, substances and products in contact with potable water shall comply in all respects, for England, with the Water Supply (Water Quality) Regulations; for Scotland, with the Water Supply (Water Quality) (Scotland) Regulations; for Northern Ireland, with the Water Supply (Water Quality) Regulations (Northern Ireland); and for Wales, with the Water Supply (Water Quality) Regulations (Wales).

2. Where an European Standard (EN) exists in respect of any chemicals used in connection with the preparation or conveyance of public water supplies, the provisions shall be complied with.

2.2 ACRYLONITRILE-BUTADIENE-STYRENE (ABS) PIPES AND FITTINGS

1. Acrylonitrile-butadiene-styrene (ABS) pressure pipes and fittings shall comply with the relevant provisions of BS 5391-1 and BS 5392-1, respectively.

2.3 ADMIXTURES FOR CONCRETE OR GROUT

(i) Clause 5.2.1 of BS EN 206-1 prohibits the use of air-entraining admixtures in standardised prescribed concrete. BS 8500-2 restricts the use of admixtures in designated concrete.

(ii) The use of accelerators, water retaining agents, etc., is not good practice in bedding to manhole covers.

1. Admixtures for use in concrete or grout shall comply with the relevant provisions of BS EN 934-2.

2. Admixtures containing chlorides shall not be used in reinforced concrete designed to retain an aqueous liquid.

2.4 AGGREGATES FOR CONCRETE

(i) Any restrictions on the source, type or group classification of aggregates should be described in the Contract.

1. Aggregates for concrete shall comply with the relevant provisions of the appropriate Standard, as set out below:

(ii) Concrete designed to retain an aqueous liquid should be described as such in the Contract.

(iii) Aggregates should also comply with the requirements of Clause 4.3 of BS 8500-2:2006.

iv) For the limitation of total chloride ion content of the concrete mix see Clause 2.25. British Standards no longer contain specific chloride limits for aggregates.

Type of Aggregate	Designated	Designed	Standardised Prescribed	Prescribed
Natural aggregates Air-cooled blastfurnace slag aggregate and other manufactured, normal weight aggregate	BS EN 12620 and PD 6682-1	BS EN 12620 and PD 6682-1	BS EN 12620 and PD 6682-1	BS EN 12620 and PD 6682-1
Lightweight aggregates		BS EN 13055-1 and PD 6682-4		BS EN 13055-1 and PD 6682-4
Coarse, recycled concrete aggregate	BS 8500-2	BS 8500-2		BS 8500-2
Coarse, recycled aggregates	BS 8500-2	BS 8500-2		BS 8500-2

2. The water absorption of aggregates for concrete designed to retain an aqueous liquid shall not exceed 3% when measured in accordance with BS EN 1097-6.

3. The proportion of coarse, recycled aggregate and coarse, recycled concrete aggregate shall not exceed 20% by mass of the total coarse aggregate in concrete. Fine, recycled aggregate or fine, recycled concrete aggregate shall not be used in concrete.

4. Recycled aggregates and recycled, concrete aggregates shall only be used in contact with raw or potable water where it is has been demonstrated that they are suitable for this application.

2.5 AGGREGATES FOR HIGH STRENGTH CONCRETE WEARING SCREEDS

1. Aggregates for high strength concrete wearing screeds (granolithic finish) shall comply with BS EN 12620 and PD 6682-1, and be 10 mm nominal size, graded in accordance with Table E.1 of PD 6682-1:2009 and shall have a Los Angeles Coefficient of no greater than 30.

2.6 AGGREGATES FOR MORTAR

1. Fine aggregates for mortar shall be washed natural sand or crushed natural stone and shall comply with BS EN 13139 and PD 6682-3.

2.7 ASPHALT CONCRETE

1. Asphalt concrete and porous asphalt for roads and other paved surfaces shall comply with the relevant provisions of BS EN 13108-1 and BS EN 13108-7, respectively.

2. The binder shall be bitumen containing no cut back agent.

3. Wearing course shall have an aggregate abrasion value (AAV) of 16 and a minimum polished stone value (PSV) of 50. Limestone aggregate shall not be used in a wearing course.

2.8 BIOLOGICAL PERCOLATING FILTER MEDIA

(i) The type of media and its nominal size and grading should be described in the Contract.

(ii) Appendix E6 of BS 1438:2004 gives recommendations for the placing of filter media.

(iii) An additional specification will be required for a filter media other than those covered by BS 1438.

1. Media for use in biological percolating filters from natural sources or derived from the reduction of iron ore in a blastfurnace, shall comply with the relevant provisions of BS 1438. All media for filter beds shall be washed, graded and screened.

2. Grading of crushed rock (e.g., granite) or blastfurnace slag filter media shall be either 75-40 mm or 63-40 mm. A further screening at 20 mm mesh size shall be undertaken on Site as the media is placed. Material passing through this screen shall not be used.

3. Structured modular crossflow media (PVC or polypropylene) products shall be specifically designed for wastewater treatment. Media assembled from sheets where the edges have been heat flanged shall not be used.

4. The media product shall be designed in such a way that it is:

(a) resistant to degradation from rot, fungi, and other forms of attack;

(b) chemically resistant to normal sewage, normally expected concentrations of acids, alkalis and organic compounds;

(c) self extinguishing and compounded to be totally stable under the action of UV light.

5. Media modules shall be designed to:

(a) be 100% crossflow media with no vertical or horizontal corrugations;

(b) have a specific surface area of 150 m^2/m^3 with 25,000 mixing points per cubic metre as a minimum and with a minimum 97% void to volume ratio;

(c) provide a minimum of 70% open area at the pack to pack interface. Horizontal surfaces shall not exceed a width of 2 mm. The media shall not allow a drop of water to free fall more than 150 mm;

(d) be capable of redistributing wastewater horizontally a minimum of 300 mm per 300 mm of media depth;

(e) allow the media to be structurally self-supporting with a minimum bearing capacity of 1.5 kN/m^2.

2.9 BITUMEN ROAD EMULSIONS

1. Bitumen road emulsions shall comply with the relevant provisions of BS 434-1.

2.10 BITUMINOUS COATINGS

1. Bituminous coatings shall comply with BS 3416.

2.11 BITUMINOUS JOINTING STRIP

1. Jointing strip for precast concrete tunnel and shaft segments shall be bitumen based, 3 mm minimum thickness and be suitable for the size and type of segments with which it is to be used. Where bolt holes are required, they shall accurately match, in size and position, the corresponding holes in the segments.

2.12 BOARDS FOR PANELLING

(i) The required type, grade and thickness of board should be described in the Contract.

(ii) Rigid urethane foam boards should be CFC free.

1. Boards for panelling shall comply with the relevant provisions of the appropriate Standard, as set out in the following table:

Board Type	Standard
Fibre building board – hard board Fibre building board – medium board Fibre building board – soft board Fibre building board – MDF	BS EN 622-1 and -2 BS EN 622-1 and -3 BS EN 622-1 and -4 BS EN 622-1 and -5
Gypsum plasterboard	BS EN 520
Expanded polystyrene boards - expandable beads	BS 3837-1
Expanded polystyrene boards – extruded boards	BS EN 13164
Rigid urethane foam - laminated board for building purposes	BS 4841-1
Rigid urethane foam - laminated board for use as a wall and ceiling insulation	BS 4841-2
Decorated laminated plastics sheet veneered boards and panels	BS 4965
Particleboard	BS EN 634-1 and -2

2.13 BOND BREAKING COMPOUND FOR DOWEL BARS

1. For water-retaining structures, bond breaking compound for dowel bars shall not contain materials likely to adversely affect potable water contained within the structure. The compound shall be constituted such that it will perform effectively for the design life of the concrete. It shall in no way retard or otherwise affect the setting of concrete.

2.14 BRICKS AND BLOCKS

(i) The type, class and grade of all new and recycled bricks and blocks should be described in the Contract.

(ii) Particular requirements for air bricks and gratings from the options listed in Appendix A of BS 493:1995 should be described in the Contract.

(iii) Bricks used in bedding to manhole covers should be bonded using high strength mortar (such as polyester resin).

1. Clay bricks, calcium silicate bricks and precast concrete masonry units shall comply with the relevant provisions of BS EN 771-1, BS EN 771-2 and BS EN 771-3, respectively.

2. Bricks to be used for manholes and chambers shall be solid Class B engineering bricks to BS EN 771-1.

3. The shapes and dimensions of special bricks shall comply with the relevant provisions of BS 4729.

4. Air bricks and gratings for wall ventilation shall comply with the relevant provisions of BS 493 and shall match any surrounding bricks.

5. Precast concrete blocks to be used below damp-proof courses shall comply with Table 12 in BS 5628-3:2005.

6. All bricks shall have freeze/thaw designation F2.

7. All bricks shall have active soluable salts content designation S2.

2.15 CAST STONE

(i) The type, constituent material and colour of cast stone should be described in the Contract.

1. Cast stone shall comply with the relevant provisions of BS 1217.

2. Reconstructed stone masonry units shall comply with the relevant provisions of BS 1217 and BS EN 771-5.

2.16 CEMENT

(i) The permitted type(s) of cement should be described in the Contract.

1. Cement shall either:

(a) be factory-produced by the cement manufacturer and comply with the provisions of the appropriate Standard, as set out below.

or

(b) be combinations, complying with BS 8500-1 Annex A, of CEM 1 cement conforming to BS EN 197-1 and fly ash conforming to BS EN 450-1 or blastfurnace slag conforming to BS EN 15167-1.

(ii) PFA (pulverised fly ash) in Portland-fly ash cement and Pozzolanic cement and in combinations, should be siliceous only. Pozzolana used in Portland-pozzolana cement should be metakaolin only.

(iii) Cement for annulus grouts should comply with either BS EN 197-1 (CEM 1, 42.5 or 52.5) or BS 4027.

Cement Type	Standard
Portland cement (CEM I)	BS EN 197-1
Portland-silica fume cement (CEM II/A-D)	BS EN 197-1
Portland-limestone cement (CEM II/A-L; CEM II/A-LL)	BS EN 197-1
Portland-pozzolana cement (CEM II/A-Q; CEM II/A-Q)	BS EN 197-1
Portland-slag cement (CEM II/A-S; CEM II/B-S)	BS EN 197-1
Portland-fly ash cement (CEM II/A-V; CEM II/B-V)	BS EN 197-1
Blastfurnace cement (CEM III/A; CEM III/B)	BS EN 197-1 or -4
Pozzolanic cement (CEM IV/B)	BS EN 197-1
Sulphate-resisting Portland cement	BS 4027
Pozzolanic pulverised-fuel ash cement	BS 6610
Masonry cement	BS EN 413-1

2. White and coloured Portland cement shall comply with the chemical and physical requirements of BS EN 197-1. Added pigments shall comply with BS EN 12878 and shall be mixed with the cement in accordance with the manufacturer's instructions. The amount of added pigments shall not exceed 10% of cement by mass, except for carbon black where the limit shall be 2%.

3. Blastfurnace slag cement for the in-situ lining of water mains shall comply with WIS 4-13-01.

2.17 CEMENT GROUTS

(i) The required class of grout, together with the type of cement and any admixture should be described in the Contract.

1. Cement grout shall be mixed in the relevant proportions indicated in the following table, using the minimum quantity of water to ensure the necessary fluidity and to render it capable of penetrating the work.

Class	Nominal Mix by Mass		
	Cement	Sand	Pfa
G1	1	-	-
G2	1	3	-
G3	1	10	-
G4	1	-	10
G5	1	-	4
G6	1	-	0.5

2. Cement grout shall be used within one hour of mixing, except where containing a grout retardant admixture.

3. Cement for fixing inserts and below plant bases shall include an approved expanding additive.

2.18 CLAY PUDDLE

1. Clay puddle shall be impervious to water and be free from sand, grit, stones and other deleterious matter.

2. The clay on being dug shall be exposed to the air for at least 24 hours and, thereafter, shall be worked with water into a consistency suitable for punning. A roll of clay, 300 mm long and 40 mm in diameter, shall support its own weight when suspended vertically from one end.

2.19 COMPRESSIBLE FILLER AND PACKING FOR PIPELINES

1. Compressible filler for interrupting concrete protection to pipelines shall consist of bitumen-impregnated insulating board to BS EN 622-1 and BS EN 622-4. The thickness of compressible filler shall be as follows:

Nominal Bore of Pipe (mm)	Thickness of Compressible Filler (mm)
Less than 450	18
450 – 1200	36
Exceeding 1200	54

2. Compressible packing for use between pipes and precast concrete setting blocks shall consist of bitumen damp-proof sheeting, complying with BS 6398.

3. Bituminous materials shall not be put into contact with plastics pipes.

2.20 CONCRETE – GENERAL

(i) The characteristic compressive strength described in the Contract should be selected, where possible, to achieve the required minimum cementitious content. If the relevant information is not available, initial testing may be required to establish the required characteristic strength. If aggregates to be used are too weak to produce the characteristic strength needed for high grade concrete, an alternative basis of assessing the cementitious content may be required.

1. Concrete shall be produced, transported and accessed in accordance with BS 8500-2.

2. The cementitious content of concrete designed to retain an aqueous liquid shall be between 325-400 kg/m^3 with a maximum free water/ cementitious ratio of 0.55.

3. The maximum size of aggregate in any structural member shall not exceed 25% of the minimum thickness of the member.

4. Conformity control and criteria shall be undertaken in accordance with BS EN 206-1:2000 Clause 8.

5. Where identity testing is specified as defined in BS 8500-1, Annex B for strength, it shall be undertaken in accordance with BS EN 206-1:2000, Annex B as follows:

(ii) Where it is important that concrete should attain its maximum impermeability within 6 months of placing, a maximum free water/ cementitious ratio less than 0.6 would be required and should be described in the Contract.

Type of Structure	Sample to Represent a Volume of (m^3)
Critical structures	10
Intermediate structures	50
Heavy concrete construction	100

(iii) See Clause 2.21 for concrete mixes containing PFA or GGBS.

(iv) Where the Contract requires aggregates of nominal size other than 20 mm to be used, the minimum cementitious content should be modified in accordance with BS 8500-1:2006 Table A7.

(v) Where there is the likelihood of unacceptable damage from alkali- silica reaction, specific precautions to minimise it should be described in the Contract in accordance with the recommendations of BRE Digest 330. Guidance relating to alkali-silica reaction is contained in BS 8500-2:2006 Section 5.2.

(vi) Examples of the type of structure in 2.20.5 are:

Critical structures:

- cantilevers, columns, suspended slabs;

Intermediate structures:

- beams, ground slabs, bridge decks, walls;

Heavy concrete construction:

- foundations, solid rafts.

2.21 CONCRETE CONTAINING PFA OR GGBS

1. For concrete mixes containing PFA or GGBS, the proportion of replacement shall be set out as in the table below:

(i) Normal combinations of cement can be used in suitable concrete to provide resistance to sulphate attack under Design Chemical classes DC-1 and DC-2. Special combinations can be used in suitable concrete to provide resistance to Design Chemical classes 3 and 4 using additional protective measures as appropiate. See Table A.6 and A.11 of BS 8500-1:2006. For further guidance see BRE Special Digest No.1 'Concrete In Aggressive Ground'.

Cementitious Component Other than Cement	Use of Concrete	Percentage by Mass of Cementitious Content	
		Normal	Special
PFA	Any	15 – 35	25 – 40
GGBS	Concrete designed to retain an aqueous liquid	0 - 50	70 - 90
GGBS	Other	0 - 65	70 - 90

(ii) Where high replacement levels are used, special care will be required in striking formwork and with curing.

(iii) The use of concrete mixes with high replacement levels may not be appropriate in areas subject to abrasion.

2. For concrete designed to retain an aqueous liquid, the maximum cementitious content for mixes containing PFA shall be 450 kg/m³ and the maximum free water:cementitious ratio shall be 0.50.

2.22 CONCRETE - READY-MIXED

(i) Any restrictions on the use of ready-mixed concrete in the Works should be described in the Contract.

(ii) A coded description for the concrete, to obviate unnecessary repetition of items upon delivery tickets, may be acceptable.

(iii) Any requirements for infor- mation concerning the taking of test cubes or slump or other consistence determinations, should be described in the Contract.

(iv) Any restrictions on the use of admixtures in the Works should be described in the Contract.

1. Where concrete is to be obtained from a ready-mix supplier, the supplying plant shall be approved by a Third Party Certification body accredited by the United Kingdom Accreditation Service (UKAS) for product conformity.

2. The delivery ticket required for each load of ready-mixed concrete shall, in addition to the information prescribed under BS EN 206-1:2000 Clause 7.3, detail:

(a) the type of aggregate;

(b) the actual cementitious content and the percentage of any PFA or GGBS included; and

(c) the position of the concrete in the Works (details to be inserted at the point of discharge).

3. All delivery tickets shall be kept at the Site and shall be made available for inspection

2.23 CONCRETE - POROUS NO-FINES

1. Porous no-fines concrete shall contain CEM 1 cement to BS EN 197-1, and 20-10 mm aggregate complying with BS EN 12620 and PD 6682-1, in a proportion of 1:10 by mass.

2. The concrete shall be mixed to a uniform colour and consistency, with the addition of water sufficient only to coat all of the aggregate without forming excess grout.

2.24 CONCRETE - AIR-ENTRAINED

(i) Air-entraining admixtures used in conjunction with PFA can give rise to very variable air content.

1. Air-entrained concrete shall have a minimum air content by volume of the fresh concrete at the time of placing, in accordance with BS 8500-1:2006 Table A.8.

2.25 CONCRETE - CHLORIDE CONTENT

1. Chloride content of fresh concrete shall be determined in accordance with BS EN 206-1:2000 Clause 5.2.7.

2. Methods for determining the chloride contents of constituent materials shall be in accordance with BS 8500-2.

2.26 CONCRETE - PIPES AND FITTINGS

(i) Particular requirements from the options listed in BS EN 1916 and BS 5911-1 should be described in the Contract.

(ii) Additional protective measures required by BRE Special Digest No.1 to provide resistance to the actual ACEC (Aggressive Chemical Environment for Concrete class) should be described in the Contract.

1. Unreinforced and reinforced concrete pipes and fittings shall comply with the relevant provisions of BS EN 1916 and BS 5911-1.

2. All pipes and fittings shall have gasket-type joints of spigot and socket or rebated form.

3. Unreinforced and reinforced concrete jacking pipes shall comply with the relevant provisions of BS EN 1916 and BS 5911-1. Pipes shall withstand the jacking loads to which they will be subjected during installation, without cracking or spalling. A certificate shall be supplied to the Client, confirming that the pipes are suitable for jacking and stating the distributed jacking loads for which they were designed.

2.27 CONNECTORS FOR TIMBER

1. Connectors for timber shall comply with the relevant provisions of BS EN 912.

2.28 COPING UNITS

(i) The required type of cement for cast coping units should be described in the Contract.

(ii) In areas of high air pollution, Type A slate coping units may be required.

1. Precast concrete, cast stone, clayware, natural stone and slate coping units shall comply with the relevant provisions of BS 5642-2. Slate coping units shall be Type B.

2.29 COPPER PIPES AND FITTINGS

1. Copper pipe for use with potable water shall comply with the relevant provisions of BS EN 1057.

2. Copper and copper alloy compression fittings for polyethylene pipes shall comply with the relevant provisions of BS EN 1254-3.

2.30 COVER BLOCKS AND SPACERS FOR REINFORCEMENT

1. Cover blocks and spacers shall comply with BS 7973-1. To maintain the correct clear cover of concrete over steel reinforcement they shall be as small as possible, consistent with their purpose.

2. Concrete cover blocks shall be manufactured with a 10 mm maximum aggregate size and otherwise produced to the same specification as the surrounding concrete. Wire cast in the block for the purpose of tying it to the reinforcement shall comply with Clause 2.128.

3. Spacers shall be of rust-proof material and shall not produce staining or otherwise be detrimental to the concrete or steel.

2.31 DAMP-PROOF COURSES

(i) Detailed requirements for damp-proof courses should be described in the Contract.

1. Damp-proof course materials shall comply with the requirements of the Standards, as set out below:

Material	Standard
Bitumen	BS 6398
Polyethylene	BS 6515
Other	BS 743

2.32 DOORS, FRAMES AND LININGS

(i) Detailed requirements for doors should be described in the Contract.

(ii) Particular requirements for wood door frames and linings (except those for fire-check flush doors) should be described in the Contract.

1. Dimensions and tolerances for wood doorsets, door leaves and frames shall be in accordance with the relevant provisions of BS 4787-1. Matchboarded doors shall comply with the relevant provisions of BS 459.

2. Dimensions and tolerances for non-wood doorsets, door leaves and frames shall be in accordance with the relevant provisions of BS 4787.

3. Steel doors in wooden frames shall comply with BS 1245.

2.33 DOWEL BARS

(i) Stainless steel should be specified in the Contract when chlorides are present.

1. Dowel bars for expansion joints in concrete shall consist of mild steel complying with the provisions of BS EN 13877-3.

2.34 DRAW CORD

1. Draw cord for duct threading shall be 8 mm diameter, 3-strand hawser laid polypropylene rope, complying with BS EN ISO 9554 and BS EN ISO 1346.

2.35 DRESSED NATURAL STONE KERBS, CHANNELS, QUADRANTS AND SETTS

(i) The required size and type of setts should be described in the Contract.

1. New dressed granite and whinstone kerbs, channels, quadrants and setts shall comply with the relevant provisions of BS EN 1342 and BS EN 1343.

2. Second-hand stone kerbs, channels, quadrants and setts shall not be excessively weathered, worn or chipped and shall be free from all bedding and jointing materials. Faces which are to remain exposed on completion shall be free from stains.

2.36 DUCTILE IRON, CAST IRON AND STEEL PIPES, FLANGES AND FITTINGS

(i) IGN 4-21-01 deals with ductile iron pipes and fittings.

(ii) The type and class of pipes, joints, fittings, gaskets, nuts, bolts and washers should be detailed in the Contract.

(iii) The required grade and thickness of steel, together with the type and strength of pipe, should be described in the Contract.

(iv) Dimensions are specified in BS EN 545 for integral flanges on ductile iron pipe. Flange facings can either be flat or raised face. Flanges in accordance with BS EN 545 and BS EN 598 are dimensionally compatible with BS EN 1092-1 and BS EN 1092-2.

(v) The nominal bar pressure rating for the pipes and flanges should be selected to meet the duty requirements and should be described in the Contract.

(vi) The type of anti-corrosion (barrier) coatings for ductile iron pipes and fittings should be detailed in the Contract, taking into account the requirements of BS EN 545 and BS EN 598. In the UK, three systems of external corrosion protection are employed: zinc spray with bituminous finishing layer with or without overwrapping of polyethylene sleeving, or zinc (or zinc/aluminium) spray with epoxy finishing layer. IGN 4-51-01 deals with external zinc coating of ductile iron pipe which is required on all pipes in the diameter range 80-1600 mm. WIS 4-52-01 and IGN 4-52-02 deal with polymeric anti-corrosion (barrier) coatings.

(vii) Any additional external protection for pipes laid in contaminated land should be described in the Contract.

(viii) Any requirement for coatings to steel pipes should be described in the Contract.

(ix) The thickness of polymeric anti-corrosion (barrier) coatings should be such that the chemical resistance test in BS EN 598:2007 Section 5.6 is met.

1. Ductile iron pipes, fittings and joints shall comply with BS EN 545 for potable water pipelines, BS EN 598 for sewerage applications and BS EN 969 for gas.

2. Cast iron pipes, fittings and joints for rainwater and soil drainage shall comply with BS EN 877.

3. Steel pipes and fittings shall conform to BS EN 10224-L275 Option 1:S or SAW. Joints shall conform to BS EN 10311.

4. Flanges for pipes and pipeline fittings shall comply with BS EN 1092-1 for steel and BS EN 1092-2 for ductile and cast iron.

5. Factory-applied coatings shall be in accordance with BS ISO 8179-1. Where external zinc spray and a bituminous finishing layer are applied, this shall be in accordance with BS EN 545.

6. Factory-applied and Site-applied (tubular) polyethylene sleeving shall be in accordance with BS 6076.

7. Cement mortar linings for potable water pipelines shall comply with the requirements of BS EN 545. Where a seal coat is required, it shall comply with the requirements of BS ISO 16132 and the complete system shall be approved under the Water Supply (Water Quality) Regulations 2000. In addition, the Instructions for Use issued by the Drinking Water Inspectorate as part of the approval shall be followed.

8. Tubular polyethylene film for use as a loose protective sleeving for buried iron pipes and fittings shall comply with the relevant provisions of BS 6076, except that the nominal layflat width shall be 280 mm for use with 80 mm and 100 mm nominal internal diameter pipelines incorporating push-in flexible joints, and 400 mm for 150 mm nominal internal diameter pipelines. Sleeving for pipes for below-ground use for potable water shall be coloured blue and all other sleeving black. Joints in sleeving shall be taped so as to form a continuous barrier and any damage to the sleeving shall be repaired prior to backfilling.

9. Bituminous coatings shall comply with BS 3416.

(x) In the UK, ductile iron pipes lined with cement mortar may require an epoxy or acrylic seal coat over the cement to prevent lime leaching with certain potable waters. This should be described in the Contract.

(xi) The use of polyethylene factory-applied sleeving as an alternative to loose sleeving should be stated in the Contract. IGN 4-50-03 gives operational guidelines for the transportation, handling and laying of ductile iron pipes with factory-applied polyethylene sleeving.

2.37 ELECTRODES, FILLER RODS AND WIRES FOR WELDING

(i) BS EN 1011-3 applies to the arc welding of stainless steel, but not to the welding of stainless steel tubes or to sections greater than 6 mm thick.

1. Electrodes, filler rods and wires for welding shall be compatible with the grade of steel to be welded.

2. Electrodes for the manual metal-arc welding of carbon and carbon manganese steel and stainless steel shall comply with the relevant provisions of BS EN ISO 2560 and BS EN 1600, respectively.

3. Electrode wires and fluxes for the submerged arc welding of carbon steel and medium tensile steel shall comply with the relevant provisions of BS EN ISO 14171.

4. Filler rods and wires for gas-shielded arc welding shall conform to the Standards in the table below:

Metal	Standard
Non alloy and fine grained steels	BS EN ISO 14341
Stainless and heat resisting steels	BS EN ISO 14343
Aluminium and aluminium alloys	BS EN ISO 18273

5. Manual welding of stainless steel shall be by the inert-gas tungsten-arc process.

2.38 EXPANDED METAL ANGLE BEADS

1. Expanded metal angle beads shall conform with BS EN 13658-1 for internal plastering and BS EN 13658-2 for external rendering.

2.39 FERTILISER

(i) The general purpose compound should be used as a single application, prior to seeding, where good root establishment but slow growth is required. The pre-seeding and post-establishment compounds require application both before and after germination, respectively, and will promote good root establishment followed by rapid growth.

1. Fertilisers shall consist of compounds containing urea nitrogen, phosphoric acid and potash in the proportions by mass, as set out below:

Chemical	General Purpose	Pre-seeding	Post-establishment
Urea nitrogen	5%	-	46%
Phosphoric acid	15%	21%	-
Potash	15%	12%	-

(ii) The compounds are not intended for use on agricultural land, where the farmer's requirements should be ascertained.

2.40 FIELD GATES

(i) Fittings and dimensions for gates should be described in the Contract.

1. Steel field gates, fittings and posts shall be hot dip galvanised in accordance with BS EN ISO 1461.

2. Timber field gates and posts shall comply with the requirements of BS 3470.

2.41 FIXING ACCESSORIES FOR BUILDING PURPOSES

1. Fixings for sheet, roof and wall coverings shall comply with the relevant provisions of BS 1494-1.

2.42 FIXINGS FOR METALWORK

(i) Sizes and types of fixings should be described in the Contract, together with minimum requirements for edge distances, centres of fixings and embedments.

(ii) For guidance on the selection and use of fixings in concrete and masonry, see CIRIA Technical Note TN137.

(iii) BS 6180 gives recommendations for fixing protective barriers.

(iv) BS 5080-1 does not give recommendations on the inter- pretation of the results of tests for the purposes of design, selection or use of fixings. CIRIA Technical Note TN75 deals with loading tests on fixings in concrete.

(v) Where described in the Contract, axial and shear loading tests on structural fixings in concrete or masonry should be carried out in accordance with the provisions of BS 5080-1 and BS 5080-2, respectively. The safe working load should be as described in the Contract.

1. Mild steel bolts and nuts shall be hot dip galvanised in accordance with BS EN ISO 1461. Mild and high tensile steel proprietary fixings shall be protected in accordance with the relevant provisions of the appropriate Standard, as set out below:

Type of Protection	Type of Fitting		
	Cast-in, Having No Machined Thread	Cast-in/Expanding, Basic Major Diameter of Machined Thread	
Hot dip galvanised	BS EN ISO 1461	-	-
Electroplated zinc	BS EN ISO 2081 Zn 10	BS 7371-12	BS EN ISO 2081 Zn 10

2. Stainless steel proprietary fixings, bolts and nuts shall be manufactured from Grade X5 CrNiMo 17-12-1 steel complying with BS EN 10088.

3. Where fixings are metallurgically incompatible with the material being fixed, suitable isolating washers and sleeves shall be used.

2.43 FLASHINGS

(i) Any requirement for a different material or thickness of sheet should be described in the Contract.

1. Flashings shall comprise milled lead strip complying with BS EN 12588.

2.44 FLEXIBLE COUPLINGS

1. Flexible couplings for gravity sewerage and drainage pipes shall comply with the provisions of WIS 4-41-01 and BS EN 295-4.

2.45 FLOOR TILES

(i) Facial sizes, thickness, colour and, in the case of floor quarries, category, should be described in the Contract.

1. Floor tiles shall comply with the relevant provisions of the appropriate Standard, as set out in the following table:

(ii) Requirements for slip resistance should be described in the Contract.

Type	Standard
Ceramic	BS EN 14411
Thermoplastics	BS EN 649
Semi-flexible PVC	BS EN 654
Unbacked flexible PVC	BS EN 649
Terrazzo – internal use	BS EN 13748-1
Terrazzo – external use	BS EN 13748-2

2.46 FOAMED CONCRETE

(i) The Contract should indicate where foamed concrete may be used in reinstatement.

1. Foamed concrete used in the reinstatement of roads, streets and highways shall comply with the requirements of the relevant Highway Reinstatement Specification.

2.47 FOAM SWABS

(i) The ratio of swab length to diameter and the diameter of the swabs for different pipe sizes should be detailed in the Contract.

(ii) The swab density and coarseness should be detailed in the Contract for each application.

1. Swabs for cleaning water mains shall be solid, cylindrical polyurethane foam, in accordance with BS EN ISO 5999.

2.48 GABIONS AND ROCKFILL MATTRESSES

(i) The design of the units (e.g., weld or woven mesh divided by partition panels), including additional protection, should be stated in the Contract.

(ii) The method, type and grading of the fill material should be stated in the Contract.

1. All gabion boxes and mattresses shall be formed from 4 mm steel wire. All wire shall be in accordance with BS 1052 and hot dip galvanised with a zinc coating to BS EN 10244-2.

2. Units shall be filled with hard, durable and clean rock, The stone shall be well graded between 100 mm and 150 mm. The units shall be assembled and filled in accordance with the manufacturer's instructions.

2.49 GASKETS FOR FLANGED AND PUSH-FITJOINTS

(i) Requirements for full face or inside-bolt-circle type gaskets should be described in the Contract.

(ii) Gasket material and type of ring should be selected for the intended duty and be stated in the Contract.

1. Gaskets for flanged pipe joints shall be either of the inside-bolt-circle type or full face type.

2. Gaskets shall be manufactured from material complying with the provisions of BS EN 681-1 or BS EN 681-2, as appropriate.

3. For potable water, gaskets shall be Type WA with a hardness range of 76-84.

4. For sewerage, gaskets shall be Type WC or WG with a hardness range of 70-90.

2.50 GENERAL FILLING MATERIALS

(i) BRE Digest 276 deals with hardcore.

1. Hardcore shall consist of clean, hard, durable material, uniformly graded from 200 mm to 20 mm and be free from extraneous matter.

2. Selected fill, whether from locally excavated material or imported, shall consist of uniform, readily-compactible material. Fill shall be free from vegetable matter, building rubbish and frozen material or materials susceptible to spontaneous combustion. It shall exclude clay of liquid limit greater than 80 and/or plastic limit greater than 55, and materials of excessively high moisture content. Clay lumps and stones retained on 75 mm and 37.5 mm sieves, respectively, shall be excluded from the fill material.

3. Quarry scalpings shall consist of hardstone passing a 38 mm screen and shall contain sufficient dust to enable the material to bind together when consolidated with water and rolled. The clay fraction shall not exceed 10%.

2.51 GLASS FOR GLAZING

(i) BS 952-2 deals with terminology for work on glass.

(ii) Requirements for glass should be described in the Contract.

1. Glass for glazing shall comply with the relevant provisions of BS 952-1.

2.52 GLASS REINFORCED PLASTICS (GRP) PIPES AND FITTINGS

(i) The pressure rating and loading of pipes and fittings should be stated in the Contract.

1. Glass reinforced plastics (GRP) pipes and fittings shall comply with the relevant provisions of BS EN 14364 for sewage and drainage applications or BS EN 1796 for water supply.

2.53 GLASS REINFORCED PLASTICS PRODUCTS

(i) Finish colour for GRP products should be stated in the Contract.

1. GRP vessels and tanks shall be designed and manufactured in accordance with BS EN 13923 and BS EN 13121-3.

2. GRP pultruded profiles shall be manufactured in accordance with BS EN 13706-2 and BS EN 13706-3. The material shall be a composite of thermosetting resin reinforced to a minimum content of 25% with continuous glass fibre complying with BS EN 14020-2 and BS EN 14020-3, with an equal number of layers in each direction. The material shall incorporate UV inhibition. The glass content shall not exceed 40% by weight. Orthophthalic polyester shall not be used on wastewater applications in aggressive environments or where frequent handling is likely.

3. Finished surfaces shall be smooth and uniform, and fibres shall be fully embedded. The top layer of reinforcement shall have sufficient resin cover to provide maximum stiffness and prevent resin chipping from unreinforced edges. There shall be no evidence of fibre orientation irregularities, inter-laminar voids, resin-rich or resin-starved areas. Fabrications shall be protected and sealed against penetration of water and dirt. Cut edges of mouldings or pultruded sections shall be fully sealed with a suitably compatible resin. The finished material shall have a minimum Barcol hardness of 35 when tested in accordance with BS 2782-10.

2.54 GLAZING MATERIALS

(i) Special requirements may be necessary for putty or other materials for use with double glazing.

1. Linseed oil putty for use in glazing wooden frames shall comply with BS 544.

2. Material for use in glazing metal frames shall be a non-setting, synthetic compound of oils, plasticisers and polymers, for gun, knife or strip application, and shall contain an agent to produce a surface skin to minimise dirt retention.

2.55 GRANULAR SUB-BASE MATERIAL

(i) The use of recycled material should be encouraged and should be described in the Contract.

1. Granular sub-base material shall be natural sands, gravels, crushed rock, crushed slag, crushed concrete or well burnt non-plastic shale. The material shall be well graded and lie within the following grading limits:

(ii) Recycled material for use in sub-bases is defined in the Highway Agency's 'Specification for Highway Works', Series 800 Clause 807 "Type 4 (Asphalt Arisings) Unbound Mixture".

(iii) IGN 4-08-01 gives guidance on pipe bedding materials.

BS 410-1 and BS 410-2 Test Sieve	Percentage by Mass Passing	
	Type 1	Type 2
75 mm	100	100
37.5 mm	85 - 100	85 - 100
20 mm	60 - 100	60 - 100
10 mm	40 - 70	45 - 100
5 mm	25 - 45	25 - 85
600 μm	8 - 22	8 - 45
75 μm	0 - 10	0 - 10

2. Where granular sub-base material is to be used within 450 mm of the surface of any road, the material shall have a heave not greater than 15 mm when tested in accordance with BS 812-124.

3. Natural sands and gravels shall be permitted in Type 2 material. Natural sands up to 12% only shall be permitted in Type 1 material.

4. The particle size shall be determined by the washing and sieving method of BS EN 933-1. The material passing a 425 μm BS sieve, when tested in accordance with BS 1377-2, shall be non-plastic for Type 1 and have a Plasticity Index of less than 6 for Type 2.

5. With the exception of well burnt non-plastic shale, the material shall have a "ten per cent fines" value of 50 kN or more when tested in accordance with BS 812-111.

6. Type 1 material shall be delivered to Site with a moisture content within + 1% and – 2% of optimum as determined in accordance with BS 5835-1, and shall be protected, so as to be maintained within this range, until its incorporation into the Works.

7. Recycled materials shall comply with all relevant Standards and Clause 2.55.

8. Type 4 granular material shall be a crushed waste concrete product. The material shall lie within the grading envelope of Table 8/5 of the 'Specification for Highway Works', and not be gap graded.

2.56 GRASS SEED

(i) The required grass seed mixture should be described in the Contract.

(ii) The four mixtures given are available nationally, examples of their individual application being as follows:

1. Grass seed shall be a tested blend of named varieties, and certificates of purity and germination shall be provided. The blend shall consist of one of the following mixtures:

Mixture	Application
1	*General application which will give satisfactory germination on a wide variety of soils*
2	*Heavy soils and wet areas*
3	*Drier, less fertile soils*
4	*Soils with very low fertility, low pH or sandy soils*

(iii) All mixtures have been selected to provide slow-growing grass with a low maintenance requirement. Mixture 1 will give the most rapid cover after germination but will require more maintenance than Mixtures 2, 3 and 4.

(iv) The mixtures are not intended for use on agricultural land or other land which may have special requirements. In these cases, the owner's and/or occupier's requirements should be ascertained.

(v) For guidance on the use of grass in hydraulic engineering practice, see CIRIA Report R116.

(vi) In view of the requirements of the Wildlife and Countryside Act 1981, consideration should be given, where appropriate, to the inclusion of an approved wild flower mix with the grass seed.

Variety	Percentage by Mass			
	Mixture 1	Mixture 2	Mixture 3	Mixture 4
Dwarf Leafy Perennial Ryegrass	20 - 30	-	-	-
Smooth-stalked Meadow Grass	25 - 35	-	20 - 30	0 - 15
Rough-stalked Meadow Grass	-	15 - 25	-	-
Creeping Red Fescue	30 - 40	40 - 50	35 - 45	20 - 50
Fine Leaved Sheep's Fescue	-	-	10 - 20	10 - 40
Chewings Fescue	-	-	-	10 - 40
Browntop Bent	5 - 15	5 - 15	5 - 15	0 - 10
Crested Dogstail	-	-	5 - 15	-
Timothy	-	20 - 30	-	-
White Clover	-	-	-	0 - 10

2.57 GRUMMETS

1. Grummets shall be of gel-impregnated hemp or plastic and have a thickness before compression of not less than 10 mm. Grummets shall be a tight fit on and shall have an external diameter at least 25 mm greater than the bolts on to which they are fitted.

2.58 GULLIES AND GULLY COVER SLABS

1. Precast concrete gullies and gully cover slabs shall comply with the relevant provisions of BS 5911-6.

2. Vitrified clay gullies shall comply with the relevant provisions of BS 65.

3. PVC-U gullies shall comply with BS 4660.

2.59 GULLY COVERS, GRATINGS AND FRAMES

1. Gully covers, gratings and frames shall comply with the relevant provisions of BS EN 124, BS 7903 and 'Highways Agency Guidance Document HA 104/09'.

(i) Some Highway and Roads Authorities demand a particular slot configuration for gully gratings, depending upon the road gradient.

(ii) BS EN 124 is a minimum performance standard only.

2.60 HANDRAILS AND BALUSTERS

(i) Where stainless steel tubes are to be bent to very small radii, it may be necessary to describe their condition as GKM(S) instead of KM.

(ii) BS 6180 deals with permanent protective barriers designed to resist vehicular impacts.

(iii) BS EN ISO 14122-3 gives guidance on the construction of guardrails.

(iv) Aluminium handrails and balusters should not be used in sewers, sewer manholes, sewage pumping stations or confined spaces in sewage treatment works.

1. Protective barriers shall conform to BS 6180 with loading determined from Table NA.8 from the UK National Annex to BS EN 1991-1-1:2006.

2. Metal handrails and balusters shall be manufactured from materials conforming to the appropriate Standard as set out below:

	Standard	Grade
Mild steel – solid	BS EN 10025-2	S275
Mild steel – tubular	BS EN 10255	HFW2
Stainless steel – solid	BS EN 10088-3	1.4401
Stainless steel – tubular	BS EN 10296-2	LW17KM
Aluminium – solid	BS EN 1559-1 and BS EN 1559-4	EN AW 6082
Aluminium – tubular	BS EN 573-3	EN AW 6082

(v) Any requirements for toeboards and infill panels should be described in the Contract.

3. GRP handrails and balusters shall be manufactured from pultruded sections conforming to BS EN 13706-2 and BS EN 13706-3. The surface shall be smooth with fibres embedded and sealed against penetration from dirt and water. The Barcol hardness of the sections shall be at least 35 when tested in accordance with BS 2782-10.

4. After manufacture, mild steel handrails and balusters shall be hot dip galvanised in accordance with BS EN ISO 1461.

5. After manufacture, aluminium handrails and balusters shall be anodised in accordance with BS EN ISO 7599, Grade AA25.

2.61 HYDRANTS

(i) Fire hydrants should be compatible with the requirements of the local Fire Authority.

(ii) The required type of hydrant, dimensions of surface box frames and covers, and size, type and material of hydrant indicator plates should be described in the Contract.

(iii) Screw-down type hydrants with loose valve plungers may not permit the passage of swabs.

(iv) The direction of closure should be stated in the Contract.

1. Hydrants, surface box frames and covers shall comply with the relevant provisions of BS 750 and BS EN 14339. Fire hydrants shall be of the screw down type (Type 2) and close in a clockwise direction.

2. Hydrant box covers shall be provided with recesses for lifting keys.

3. Hydrant indicator plates shall comply with the relevant provisions of BS 3251.

4. The surfaces of all the hydrant components shall be protected from corrosion either by the nature of their material of construction or by coating in accordance with WIS 4-52-01. Internal water-wetted surfaces shall be coated to Class A standard; all other surfaces shall be coated to Class B.

5. All fasteners used in the assembly of hydrants shall be protected in accordance with Clause 2.86.

6. The frame and cover shall be Grade A to BS 750 and BS EN 14339 and have a clear opening of not less than 380 x 230 mm.

2.62 IMPORTED TOPSOIL

(i) For the definition of in-situ topsoil see Clause 3.3.

(ii) Any requirements for testing for club root and eelworm should be described in the Contract.

(iii) Attention is drawn to Section 14.2 of the Wildlife and Countryside Act 1981, which prohibits deposition of any plant material from any non-native invasive species listed in Part 2 of Schedule 9 of the Act (e.g. Giant Hogweed or Japanese Knotweed).

(iv) Organic land is defined as land cultivated without the addition of artificial fertiliser or use of pesticides or herbicides for at least the previous five years, as certified by the Soil Association, or other equivalent body.

1. Imported topsoil shall comply with BS 3882 and be of light or medium texture, having a pH value of between 6.0 and 7.5. Imported topsoil shall not contain stones greater than 20 mm in size nor have a total stone content exceeding 10% by mass.

2. Organic topsoil shall be imported from an organic land estate having a soil composition from a similar soil series. Details of the source and composition of the topsoil shall be provided, unless stated in the Contract.

2.63 IMPORTED TURF

(i) For in-situ turf for relaying, see Clauses 3.2 and 3.9.

(ii) BS 3969:1998 Clause 3 permits up to 50% of Dwarf Leafy Perennial Ryegrass.

1. Imported turf shall comply with BS 3969 and be delivered to the Site within 24 hours of lifting (18 hours in spring and summer). Constituent grasses and their proportions shall comply with the provisions of BS 3969:1998, Clause 3.

2.64 INDUSTRIAL FLOORING, WALKWAYS AND STAIR TREADS

(i) The information required by Clause 4 of the relevant Part of BS 4592 should be described in the Contract.

(ii) BS EN ISO 14122-2 and BS EN ISO 14122-3 give guidance on the construction of walkways and stairs.

(iii) The weight of galvanising should be described in the Contract to suit the resistance to the corrosiveness encountered. BS EN ISO 14713-1 gives coating thickness/weight for typical environment classifications.

(iv) Cut openings and edges of open type floor panels should be resealed to prevent water ingress. All straight edges, and where the shape allows, should be fully and neatly trimmed with a full depth binding-bar.

(v) Reference should be made to the Manual Handling Operations Regulations.

1. Open type and plate flooring shall comply with BS 4592-0 and the relevant part of BS 4592, set out in the table below:

Material and Type	BS 4592 Part
Metal open bar	1
Expanded metal	2
Cold formed metal planks	3
GRP open bar	4
Metal and GRP solid plates	5
Moulded GRP open mesh	6

2. After manufacture mild steel flooring, walkways and stair treads shall be hot dip galvanised, in accordance with BS EN ISO 1461.

3. Individual floor panels shall be bolted down to prevent unauthorised removal and shall be prevented from sliding sideways when adjacent panels are removed by means of permanent steelwork or non-removable locating studs. Panels that require frequent lifting shall be hinged and be readily lifted by one man. Removable floor panels and support beams shall be stamped with reference numbers to ensure correct replacement.

4. Where bolting is metallurgically incompatible with the material being fixed, the contact areas shall be isolated either by painting with an approved silicon sealant (and allowed to dry before tightening together) or with suitable isolating washers and sleeves.

2.65 JOINERY TIMBER

1. Timber for joinery shall comply with the relevant provisions of BS EN 942.

2. The sizes of sawn and processed softwood and hardwoods shall comply with the relevant provisions of BS EN 1313-1 and BS EN 1313-2, respectively.

2.66 JOINT FILLER BOARD

1. Filler board for joints in concrete (other than softwood) shall comply with the requirements and tests in Clause 1015 of the 'Specification for Highway Works'. Knot-free softwood, preserved in accordance with Clause 2.126, may be used for joints in concrete carriageways.

2. Holes in preformed joint filler to accommodate dowel bars shall be accurately bored or punched out to produce a sliding fit on the dowel bars.

3. The material comprising the joint filler shall be of such quality that it can be satisfactorily installed in position at the joint.

4. Adhesives used to retain preformed joint fillers in place during construction shall have no harmful effects on concrete and, except for those used in connection with softwood fillers, shall be obtained from the same manufacturer as the joint filler.

5. Preformed filler for joints in structures to retain aqueous liquids shall have a maximum water absorption of 0.3% by volume and a non-recovered compression set of 20% of the original thickness, both when tested in accordance with ASTM D3575-08.

2.67 JOINT SEALING COMPOUNDS AND SEALANTS

(i) Fuel-resistant types of sealants to BS EN 14188-1 or BS 5212-1 may be required where concrete surfaces are subject to regular fuel spillage.

(ii) There are no current British Standards for the cold-applied, non-curing and heat-softened, hand-applied types of sealant, though various such materials are available in proprietary form.

(iii) BS 6213 gives guidance on the selection of constructional sealants.

(iv) In certain circumstances, polyurethane-based sealants (for which there are no current British Standards) may be more resistant to biodegradation than poly- sulphide-based ones.

(v) CIRIA Technical Notes TN128 and TN144 deal with civil engineering sealants in wet conditions.

1. Joint sealing compounds shall be impermeable ductile materials of a type suitable for the conditions of exposure in which they are to be placed and capable of providing a durable, flexible and watertight seal by adhesion to the concrete throughout the range of joint movement.

2. Hot poured joint sealants shall comply with BS EN 14188-1 Type N1 sealant.

3. Cold poured polymer-based joint sealants shall comply with BS 5212-1 Normal Type N sealant.

4. Primers for use with joint sealants shall be compatible with, and obtained from the same manufacturer as, the adjacent sealant. Primers shall have no harmful effects on concrete.

5. Sealants and primers which will be in contact with sewage or sewage sludge shall be resistant to biodegradation.

6. Two-part polyurethane joint sealants shall comply with the requirements of BS EN ISO 11600.

2.68 JOINT SEALS AND LUBRICANTS

(i) BS EN 681-1 does not contain a microbiological test requirement for elastomeric seals.

1. Elastomeric joint seals shall be obtained from the pipe manufacturers and shall comply with the provisions of BS EN 681-1. For water mains, the seals shall be Type WA and for drainage purposes they shall be either Type WC or WG.

2. Seals shall be tested in accordance with BS 7874 and shall comply with the following:

- the average loss in mass (Z) of the test pieces shall not exceed 3.5%; and

- there shall be no greater release of carbon black or other fillers from the test set than from the control set when the surface of the specimens is lightly rubbed.

In the case of composite seals, the requirements apply only to those components exposed to the contents of the pipeline or pipework.

3. Joint lubricants for sliding joints shall have no deleterious effects on either the joint rings or pipes and shall be unaffected by the liquid to be conveyed. Lubricants to be used for jointing water mains shall not impart to water, taste, colour or any effect known to be injurious to health, and shall be resistant to bacterial growth.

2.69 JOIST HANGERS

(i) The material and protection requirements shall be selected from BS EN 845-1:2003 Table A.1 and stated in the Contract.

1. Joist hangers for building into masonry walls or the inner skin of cavity walls shall comply with the provisions of BS EN 845-1.

2.70 LADDERS

(i) BS EN ISO 14122-4 gives guidance on the construction of ladders.

(ii) The weight of the galvanising should be described in the Contract to suit the resistance to the corrosiveness encountered. BS EN ISO 14713-1 gives coating thickness/weight for typical environment classification.

1. Ladders, other than those in manholes, shall conform to BS 4211. Ladders in manholes shall conform to BS EN 14396 with width of rung of 400 mm and two stringers.

2. Mild steel ladders for vertical fixing shall be fabricated from steel conforming to BS EN 10025-2. After fabrication, low carbon steel ladders shall be hot dip galvanised in accordance with BS EN ISO 1461.

3. Stainless steel ladders for vertical fixing shall be fabricated from Grade X5CrNiMo 17-12-2 steel conforming to BS EN 10088-3.

4. Aluminium ladders for vertical fixing shall be fabricated from Grade EN AW-6082 aluminium conforming to BS EN 573-3. After fabrication they shall be anodised in accordance with BS EN ISO 7599 Grade AA25.

5. GRP ladders shall be manufactured from pultruded sections conforming to BS EN 13706-2 and BS EN 13706-3. The surface shall be smooth with fibres embedded and sealed against penetration from dirt and water. The Barcol hardness of the sections shall be at least 35 when tested in accordance with BS 2782-10.

2.71 LEAD

1. The chemical composition of all lead to be used for jointing or caulking shall comply with the provisions of BS EN 12588.

2. Lead wool shall be extruded to produce strands of triangular cross-section.

2.72 LIME FOR MORTAR

1. Lime for mortar shall be in the form of lime putty, complying with the relevant provisions of BS EN 459-1.

2.73 LINTELS

(i) The required type and size of lintel should be described in the Contract.

(ii) BS 5977-1 gives a method for assessment of the load carried by lintels in masonry.

1. Concrete, steel or timber prefabricated lintels shall comply with the relevant provisions of BS EN 845-2.

2.74 MANHOLE COVERS AND FRAMES

1. Manhole covers and frames shall comply with the relevant provisions of BS EN 124, BS 7903 and the Highway Agency's 'Chamber Tops and Gully Tops for Road Drainage and Services: Installation and Maintenance' HA 104/09.

2. All manhole covers shall have closed keyways.

3. Chamber access covers with a clear opening exceeding 1 m shall conform to BS 9124.

(i) The shape and size of the opening should be described in the Contract.

(ii) The Contract should indicate any requirements for locking devices.

(iii) The responsibilities for supplying manhole lifting keys should be detailed in the Contract.

(iv) BS EN 124 is a minimum performance standard only.

(v) The maximum weight of components of covers should be specified in the Contract. The specifier should have regard to the Manual Handling Operations Regulations.

(vi) Covers and frames should be described in the Contract.

(vii) Covers and frames to manholes containing air valves on water mains should be of a ventilated type.

(viii) The information required by BS 9124:2008 Annex A should be stated in the Contract.

2.75 MANHOLE STEPS

1. Steps for manholes and other chambers shall be Type D Class 1, complying with the requirements of BS EN 13101.

(i) For all manholes, the size and spacings and type of manhole steps should be described in the Contract.

(ii) Unprotected aluminium steps should not be used.

2.76 MARKER TAPE AND MARKER POSTS

1. Marker tape for buried water mains shall be blue PVC or polyethylene ribbon at least 150 mm wide, incorporating a corrosion-resistant tracing system and shall be printed with the words "WATER MAIN" in bold capital letters throughout its length and at intervals not exceeding 700 mm.

2. Marker tape for buried cables shall be yellow PVC or polyethylene ribbon at least 150 mm wide and shall be printed with the words "ELECTRIC CABLE" in bold capital letters throughout its length and at intervals not exceeding 700 mm.

3. Marker tape for buried wastewater pipes or pumping mains shall be red PVC or heavy gauge polyethylene sheeting at least 150 mm wide and shall be printed with the words "GRAVITY SEWER" or "PRESSURE SEWER" in bold capital letters throughout its length and at intervals not exceeding 700 mm and shall incorporate a corrosion-resistant tracing system for non-metallic pipes.

4. Valve or pipeline marker posts shall be precast reinforced concrete 1120 mm high. The tops of marker posts shall be fixed so that their finished level is 500–600 mm above the final surface.

5. Marker posts shall be fitted with a 180 x 205 mm indicator plate to indicate WO, SV, AV or FH, as appropriate, together with main size and distance from post. Plates shall comply with BS 3251 and shall be fixed to the post with stainless steel fixings.

6. The colour for fire hydrant plates shall be in accordance with BS 3251 (black letters on a yellow background).

(i) Consideration should be given to the use of a marker tape and any requirements described in the Contract.

(ii) In the case of fire hydrants, the requirements of the Fire Authority should be established and included in the Contract.

(iii) Appropriate marker tapes for other buried services, such as fibre-optic cables, should be described in the Contract.

(iv) Polyethylene mesh should only be used as a tracing system, and placed beneath the pipework to reduce the risk of future damage to the mesh from excavation.

(v) Where other types of material for marker posts are required, details should be given in the Contract.

2.77 MASTIC ASPHALT

(i) Grade IV mastic asphalt flooring is suitable for loading sheds and heavy duty factory floors. Lower grades may be appropriate in other circumstances.

(ii) Mastic asphalts with natural rock asphalt aggregates are not normally necessary.

(iii) Where an acidic environment is expected, the Mastic Asphalt Council (www.masticasphaltcouncil.co.uk) should be consulted for advice on suitability of limestone aggregates.

1. Mastic asphalt for building and civil engineering shall comply with the relevant provisions of BS 6925:1988 Table 25, as set out below:

Application	Type
Roofing	R988
Flooring	F1076, Grade IV
Tanking	T1097
Coloured flooring	F1451, Grade IV

2.78 MECHANICAL COUPLINGS FOR PIPELINES AND FITTINGS

(i) On potable water pipelines, it should be noted that all mechanical couplings and fittings are regarded as being in contact with potable water, see Clause 2.1.

(ii) The Contract should indicate the pressure ratings required.

(iii) The type of end loading restraint should be described in the Contract.

(iv) IGN 4-52-02 provides guidance on polymeric anti-corrosion (barrier) coatings and complements WIS 4-52-01.

1. Mechanical couplings and repair clamps for iron pipes shall comply with WIS 4-21-02.

2. Mechanical joints and fittings for polyethylene pipes in nominal size 90 mm or above for use with cold potable water shall comply with WIS 4-24-01. Mechanical joints and fittings for polyethylene pipes less than or equal to nominal size below 63 mm for use with cold potable water shall comply with WIS 4-32-11 or BS EN 1254-3.

3. Fittings for PE80 (MDPE) and PE100 (HDPE) shall be Type 2, fit for the purpose, end load restraining (except for unusual locations) and have an internal pipe support.

4. All fittings shall be protected against corrosion by the application of a polymeric barrier coating in accordance with WIS 4-52-01. All internal and external surfaces shall be coated to Class B as a minimum.

5. All fasteners shall be protected from corrosion by the application of zinc and a polymeric barrier coating in accordance with WIS 4-52-03.

2.79 METAL LATHING

1. Expanded metal and ribbed lathing for internal plastering and external rendering, shall comply with the relevant provisions of BS EN 13658-1 and BS EN 13658-2, respectively.

2.80 METAL TIES

(i) British Standard DD 140 gives recommendations for the design of wall ties.

1. Metal ties for cavity wall construction shall comply with the relevant provisions of BS EN 845-1.

2.81 MINERAL AGGREGATES FOR FLAT ROOFS

(i) Where an acidic environment is expected, the Mastic Asphalt Council (www.masticasphaltcouncil.co.uk) should be consulted for advice on suitability of limestone aggregates.

1. Mineral aggregates for the reflection of solar heat on flat built-up bitumen or mastic asphalt roofs shall be light coloured and consist of a hard limestone having a low moisture absorption characteristic, granite, gravel, calcined flint, calcite or felspar of 10 mm nominal size.

2.82 MORTAR

1. Mortar shall be mixed only when required and in the relevant proportions indicated in Table NA.2 of the UK national annex to BS EN 1996-1-1:2005 until its colour and consistency are uniform. The constituent materials shall be accurately gauged, allowance being made for the bulking of fine aggregate.

2. All mortar shall be conveyed fresh as required for use. Mortar which has begun to set or which has been Site-mixed for a period of more than one hour in the case of Classes M12 and M6, and two hours in the case of Classes M4 and M2 shall not be used. Plasticising and set-retarding admixtures shall conform to BS EN 934-3.

3. Ready-mixed and ready-to-use rendering and masonry mortar shall conform to BS EN 998-1 and BS EN 998-2, respectively.

4. Rapid hardening cementitious, epoxy resin or polyester resin mortar for setting manhole covers and frames shall have a minimum working time of 15 minutes. The mortar shall reach a minimum compressive strength of 30 N/mm^2 and minimum tensile strength of 5 N/mm^2 within 3 hours of mixing.

2.83 NAILS

1. Nails shall comply with the relevant provisions of the appropriate Standard, as set out below:

Type	Standard
Steel	BS 1202-1
Copper	BS 1202-2
Aluminium	BS 1202-3

2.84 NATURAL STONE

1. Natural stone shall be of durable quality, uniform in texture, and free from iron bands, spots, sandholes, flaws, shakes and other imperfections which would adversely affect its strength or appearance. The dimensions of stones shall be adequate for proper coursing and bonding.

2.85 NON-MAN ACCESS CHAMBERS

1. Non-man access chambers shall comply with the relevant provisions of BS EN 752.

2.86 NUTS, SCREWS, WASHERS AND BOLTS

(i) Any protective coatings required should be described in the Contract.

(ii) The type of nuts, screws, washers and bolts required should be described in the Contract.

(iii) The specific environmental conditions on Site would determine the grade of the material required.

1. Nuts, screws, washers and bolts shall comply with the relevant provisions of the appropriate Standard, as set out below:

Type	Standard
Metal washers for general purposes	BS 4320
ISO black bolts, screws and nuts	BS 4190
ISO precision bolts, screws and nuts	BS 3692
High-strength friction grip bolts, nuts and washers	BS 4395-1 or relevant parts of BS EN 14399
Stainless steel bolts, screws, studs	BS EN ISO 3506-1
Stainless steel nuts	BS EN ISO 3506-2

2. Bolting for pipes and fittings shall comply with the relevant provisions of BS EN 1092-2, except that spheroidal graphite iron bolts for use with ductile iron pipes and fittings shall be manufactured from metal complying with the provisions of BS EN 1563 for Grade EN-JS1050.

3. Bolt length and tightening torque shall be in accordance with manufacturer's recommendations and shall be sufficient to ensure that nuts are full-threaded when tightened in their final position with two threads showing.

4. Where bolting is metallurgically incompatible with the material being fixed, the contact areas shall be isolated either by painting with an approved silicon sealant (and allowed to dry before tightening together) or with suitable isolating washers and sleeves.

5. Washers shall be provided under the head of the bolt and under the nut.

6. Unless manufactured of stainless steel, all fasteners shall be protected against corrosion in accordance with WIS 4-52-03.

2.87 PACKINGS FOR TUNNELS

1. Packing pieces for joint gaps in bolted tunnel segments shall be band-sawn, knot-free softwood, preserved in accordance with Clause 2.126.

2.88 PAINTS AND PAINTING MATERIALS FOR BUILDINGS

(i) This Clause applies to paint work to BS 6150 - see Clause 6.26.

(ii) The required type of priming paint should be described in the Contract.

(iii) Paints for the protection of steelwork against corrosion and paints for the fire protection of steelwork should be described in the Contract.

1. Ready-mixed paints for buildings shall be of external quality. Paint colours for building purposes shall comply with the relevant provisions of BS 4800.

2. Raw, refined and boiled linseed oils for paints and varnishes shall comply with the relevant provisions of BS EN ISO 150.

3. Knotting for use as an impervious covering for knots and other resinous areas shall comply with BS 1336.

4. Stopping shall comprise a mixture of one third white lead to two thirds ordinary whiting and linseed oil putty, with a small quantity of gold size added.

5. Priming paint for wood shall comply with the appropriate Standard, as set out below:

Type	Standard
Ready-mixed aluminium	BS 4756, Type II
Water-borne	BS 7956, Type B
Solvent-borne	BS 7956, Type B

6. Priming paint for metal shall comply with the relevant provisions of BS 4652. Lead-based priming paint shall not be used.

7. Paint remover shall be non-flammable, solvent-based and comply with BS 3761.

8. All painting materials for one operation shall be compatible and shall be obtained from one manufacturer.

2.89 PERMANENT FENCING

(i) Where appropriate, the various Parts of BS 1722 provide for concrete for surrounding the bases of posts.

1. Permanent fencing shall comply with the relevant part of BS 1722, as set out in the following table:

(ii) All timber for permanent fencing should be given preservative treatment, in accordance with the provisions of the relevant Standard.

BS 1722	Type of Fencing
Part 1	Chain link fences
Part 2	Strained wire and wire mesh netting fences
Part 4	Cleft chestnut pale fences
Part 5	Close boarded and wooden palisade fences
Part 7	Wooden post and rail fences
Part 8	Mild steel (low carbon steel) continuous bar fences and hurdles
Part 9	Mild steel (low carbon steel) fences with round or square verticals and flat horizontals
Part 10	Anti-intruder fences in chain link and welded mesh
Part 11	Prefabricated wood panel fences
Part 12	Steel palisade fences
Part 14	Open mesh steel panel fences

2.90 PIPE SURROUND MATERIALS

(i) Any limitations on the size and type of materials should be described in the Contract.

(ii) IGN 4-08-01 gives guidance on pipe surround materials.

1. Processed granular and as-dug bedding, sidefill and surround materials for buried pipelines shall comply with WIS 4-08-02.

2. Recycled materials shall comply with BS 8500-2.

2.91 PIPES FOR DUCTS

(i) BS 8313 gives guidance on ducts for building services.

(ii) Certain Public Utilities may require ducts to be of a particular colour.

(iii) Grey PVC-U to BS 3506 may be used where ducts are not required to be pressure tight.

1. Pipes, joints and fittings for exposed ducts for building services shall comply with the appropriate Standard, as set out below:

Type	Standard
Unplasticised PVC pipe	BS EN ISO 1452- 2 and BS EN ISO 1452-3 or BS 3506
Hollow steel sections (greater than 150 mm OD)	BS EN 10210-2
Steel tubes (not greater than 150 mm OD)	BS EN 10296-1 or BS EN 10297-1

2. Pipes, joints and fittings for buried ducts shall have flexible mechanical joints and comply with the relevant provisions of the appropriate Standard, as set out below:

Type	Standard
Vitrified clay	BS 65 or BS EN 295-1
Unreinforced or reinforced concrete	BS EN 1916 and BS 5911-1
Unplasticised PVC	BS 4660 or BS EN 1401-1
Structured wall plastics pipes	BS EN 61386-24

3. All cable ducts shall be fitted with draw cords.

4. Pipes for cable ducts shall be coloured as follows:

Duct	Colour
Electricity power cables	Black
Electrical signal and telecoms cables	Grey
Cable TV	Green
Street Lighting	Orange
Motorway communications	Pink

2.92 PIPES FOR LAND DRAINAGE AND TEMPORARY DRAINS

1. Pipes, joints and fittings for land drainage and temporary drains shall comply with the relevant provisions of the appropriate Standard, as set out below:

Type	Standard
Vitrified clay	BS 65 or BS EN 295-1
Perforated vitrified clay	BS EN 295-5
Clayware field drains	BS 1196
Plastic field drains	BS 4962

2.93 PLASTER

1. Premixed lightweight plaster shall comply with the relevant provisions of BS EN 13279-1.

2. Polyvinyl acetate emulsion bonding agents for indoor use shall comply with BS 5270-1.

2.94 PLASTIC CHAMBERS AND RINGS

1. Plastic chambers and rings, including demarcation chambers, shall comply with BS EN 13598-1 or BS EN 13598-2.

2.95 PLASTIC SHEETING

1. Plastic sheeting for waterproof underlay shall have a composition in accordance with BS 6076.

2. The minimum nominal film thickness shall be 250 µm.

2.96 PLYWOOD

(i) The required thickness, grade and type of bonding should be described in the Contract.

(ii) Proprietary products are available for these applications.

1. Plywood for general use shall comply with the relevant provisions of BS EN 636. Marine grade plywood shall be in accordance with BS 1088-1.

2. All plywood used for formwork on water supply structures in temporary and permanent works on the side of the concrete facing the potable water shall either be phenol free or be faced with a surfacing that prevents phenol from contacting the concrete.

3. Interior quality plywood shall have MR bonding and external quality plywood shall have WBP bonding.

2.97 POLYETHYLENE PIPES AND FITTINGS

(i) For co-extruded polyethylene pipes, identification stripes may be included in the blue outer layer .

(ii) The pressure rating should be described in the Contract.

(iii) Further guidance on the design and use of PE pipes can be found in the WRc/BPF 'Polyethylene Pipe Systems Manual' and IGN 4-32-18.

(iv) The required short-term surge pressure resistance and the lifetime at a pressure of 1.2 times MRS (Minimum Required Strength as defined in BS EN 12201-1) to determine the safe duration of pressure tests should be stated in the Contract.

(v) Clause 2.78.2 covers mechanical joints and fittings for polyethylene pipes for use with cold potable water.

(vi) Pipes and cisterns conveying and holding water that is not potable should be marked or colour coded in accordance with BS 1710.

(vii) Black polyethylene pipes with blue stripes should only used for water supply if stated in the contract.

1. Polyethylene piping systems for water supply shall comply with BS EN 12201-1 and BS EN 12201-2 and be coloured blue. Co-extruded polyethylene pipes may be used for water supply but shall have a blue outer layer on top of blue or black inner layers.

2. Polyethylene piping systems for above-ground pressure systems for general purposes, for drainage and for sewerage shall comply with BS EN 12201-1 and BS EN 12201-2 and be coloured black. Co-extruded polyethylene pipes may be used for these purposes but shall have a black outer layer.

3. Polyethylene fittings for use with cold potable water shall comply with the relevant provisions of BS EN 12201-3.

4. Electrofusion fittings shall comply with the relevant provisions of BS EN 12201-3.

5. All electrofusion fittings shall be of integral wire construction. All fittings shall be of an automatic type and fitted with recognition resistors, identifiable by an automatic electrofusion control box, complete with an electronic data acquisition facility for joint data analysis and quality assurance.

2.98 PRECAST CONCRETE SLABS AND COVER FRAME SEATING RINGS

(i) Additional protective measures required by BRE Special Digest No. 1 to provide resistance to the actual ACEC (Aggressive Chemical Environment for Concrete class should be described in the Contract.

1. Precast concrete slabs and cover frame seating rings shall comply with the relevant provisions of BS EN 1917 and BS 5911-3.

2.99 PRECAST CONCRETE FLAGS AND PAVING BLOCKS

1. Precast concrete flags shall be hydraulically pressed and shall comply with the relevant provisions of BS EN 1339. Flags shall be 50 mm thick.

2. Precast concrete paving blocks shall comply with the relevant provisions of BS EN 1338.

2.100 PRECAST CONCRETE KERBS, CHANNELS, EDGINGS AND QUADRANTS

1. Precast concrete kerbs, channels and edgings shall be hydraulically pressed, and they, and precast concrete quadrants, shall comply with BS EN 1340. Where kerbs or channels are required to be laid to a radius of 12 m or less, components of the appropriate radius shall be used.

2.101 PRECAST CONCRETE MANHOLES AND SOAKAWAYS

(i) Particular requirements from the options listed in BS EN 1917 and BS 5911-3 should be described in the Contract.

1. Precast concrete manhole and soakaway units of circular cross-section shall comply with the relevant provisions of BS EN 1917 and BS 5911-3. Units which bed onto bases shall be manufactured so that imposed vertical loads are transmitted directly via the full wall thickness of the unit.

(ii) Additional protective measures required by BRE Special Digest No. 1 to provide resistance to the actual ACEC (Aggressive Chemical Environment for Concrete) class should be described in the Contract.

2. The profiles of joints between units and the underside of slabs, shall be capable of withstanding applied loadings from such slabs and spigot-ended sections shall only be used where the soffit of the slab is recessed to receive them.

3. Precast concrete chamber sections for valves and meters shall be interlocking and comply with BS EN 1917 and BS 5911-3.

2.102 PRECAST CONCRETE BOX CULVERTS

(i) The design of culverts is detailed in CIRIA Guide C698.

1. Precast concrete box culverts shall comply with BS EN 14844.

2.103 PRECAST CONCRETE SEGMENTS FOR TUNNELS AND SHAFTS

(i) The required type of cement should be described in the Contract.

(ii) A review of current test procedures for precast concrete tunnel linings can be found in CIRIA Technical Note TN104.

1. All concrete used in the manufacture of segments shall be compressive strength class C32/40 as defined in BS 8500-1.

2. Concrete shall be sampled and tested for compliance with the specified strength class in accordance with the provisions of BS EN 206-1:2000, Clause 8.2.1. The concrete shall be subject to identity checks of not less than one sample per 20 m^3 of fresh concrete. All segments shall have the date of manufacture clearly marked in an appropriate position at the time of manufacture.

3. Segments shall not be removed from the moulds until the concrete cube strength has reached 10 N/mm^2 and no segments shall leave the place of manufacture or be used until 28 days after casting. Segments shall not be incorporated until the relevant test results confirm that the concrete complies with the specified compressive strength class.

4. Segments shall be cast with such accuracy and uniformity of dimensions that all similar segments shall be interchangeable, not only within individual rings but with corresponding segments of other rings. All surfaces of the segments shall be free from cracking, honeycombing or other blemishes.

5. Segments shall be manufactured to the following tolerances:

Nominal Dimension	Permissible Deviation (mm)
Circumferential length	±1.5
Radius of curvature	±3
Thickness	±3 (on back face only)
Width	±1.5

6. Segments shall be designed to have a water absorption not exceeding 6% by mass when tested in accordance with BS EN 1917.

7. Segments shall withstand handling, erection and any shield thrust stresses, without cracking, spalling or distortion.

8. The clear cover of concrete over any steel reinforcement shall be not less than 12 mm and spacers shall be of corrosion-resistant material.

9. All segment joint faces shall have a caulking rebate of minimum size 20 mm deep by 3 mm wide for bolted segments and 10 mm deep by 3 mm wide for smooth-bore segments.

10. Where grouting is described in the Contract, all segments shall have at least one grout hole of 50 mm diameter.

11. Tapered segments for curves shall comply with the general requirements of this Clause, each segment having the radius and the location of the segment within the ring, clearly marked. Segments shall be symmetrically tapered.

2.104 PRECAST CONCRETE SETTING BLOCKS FOR PIPES

1. Precast setting blocks for pipes shall have rectangular faces, with sufficient plan area to prevent punching of the blinding concrete or final surface, and to provide an adequate seating for the pipes. They shall be manufactured from compressive strength C16/20 concrete using the same type of cement as in the adjacent concrete bed. Blocks shall not be used until they have achieved a cube strength of 13.5 N/mm^2.

2.105 PRECAST CONCRETE TANKS

(i) BS EN 1992-3 covers the design and construction of structures used for retaining aqueous liquids. Its scope includes precast concrete.

(ii) The type of seal (WA, WB, WC, WD, etc.) should be described in the Contract.

(iii) The required exposure class of concrete should be stated in the Contract.

1. The design and construction of concrete panels for pre-cast concrete tanks shall comply with BS EN 1992-3.

2. Rubber and thermoplastics seals between panels, floor slabs, etc., shall comply with BS EN 681-1 and BS EN 681-2, respectively.

3. Class C32/40 concrete shall be used as a minimum for the concrete panels.

2.106 PRESTRESSED CONCRETE PIPES AND FITTINGS

(i) Particular requirements from the options listed in BS EN 639, BS EN 642 or BS 5911-5 should be described in the Contract.

(ii) The Contract should describe whether cylinder or non-cylinder type pipes and fittings are required.

1. Prestressed concrete pressure pipes and fittings shall comply with the relevant provisions of BS EN 639, and BS EN 642. Prestressed concrete pipes and fittings for drainage and sewerage purposes shall comply with the relevant provisions of BS 5911-5.

2. Unless steam cured, no pipes or fittings shall leave the place of manufacture until they have been allowed to cure and mature under suitable conditions for a total period of not less than 28 days.

2.107 PRESTRESSED PRECAST CONCRETE FLOORS

1. Precast prestressed concrete floors shall be formed from units manufactured to BS EN 13369.

2.108 PROFILED STEEL SHEETING

(i) Details of profiled steel sheeting should be described in the Contract.

1. Profiled steel sheeting and cladding shall comply with BS EN 14782.

2.109 PROPYLENE CO-POLYMER PRESSURE PIPES

1. Propylene co-polymer pressure pipe shall comply with the relevant provisions of BS 4991 and, where it is to be in contact with potable water, shall be Series 1.

2.110 PTFE TAPE

1. Unsintered polytetrafluoroethylene (PTFE) tape for thread sealing applications shall comply with BS 7786.

2.111 PULVERISED-FUEL ASH

1. Pulverised-fuel ash (PFA) and fly ash, for use as a component material in cementitious grout or non-structural concrete, shall comply with BS 3892-2 and BS 3892-3.

2. PFA for use as a cementitious component in structural concrete and annulus grouts shall comply with BS EN 450-1 fineness category S, loss on ignition category A or B. PFA for annulus grouts shall be pre-blended and bagged before delivery to the Site.

3. Conditioned PFA or fly ash for use as a fill material shall be supplied dry and shall be compacted to ± 2% of the optimum moisture content and maximum dry density in the ranges 18-25% and 1200-1500 kg/m^3, respectively, when determined in accordance with BS 1377-4 (light or heavy Hammer Method rather than Rammer Method). The PFA fill shall not be reused when producing the compaction curve.

2.112 RAINWATER PIPES AND GUTTERS

(i) The Contract should describe the section required for aluminium and PVC-U gutters.

(ii) The Contract should describe the required grade and shape of sheet and strip aluminium pipes and gutters.

(iii) BS EN 877 deals with flexible joints for cast iron drainpipes and fittings.

(iv) The Contract should describe whether ears are required on cast iron pipes and fittings.

(v) The Contract should describe the required colour of PVC-U pipes and fittings.

1. Rainwater pipes, gutters, fixings and accessories shall comply with the relevant provisions of the appropriate Standard, as set out below:

Material	Standard
Cast iron	BS 460
Aluminium	BS EN 612
Steel	BS EN 612
PVC-U (gutters)	BS EN 607
PVC-U (downpipes)	BS EN 12200-1
Brackets for gutters to BS EN 607 and BS EN 612	BS EN 1462

2.113 ROLLED ASPHALT

1. Hot rolled asphalt shall comply with the relevant provisions of BS EN 13108-4.

2.114 ROOF COVERINGS

(i) The required type, grade, category, classification, size, group or colour of roof covering material should be described in the Contract.

1. Roof coverings shall comply with the relevant provisions of the appropriate Standard, as set out in the following Table:

Material/Type	Standard
Clay tiles and fittings	BS EN 1304
Concrete tiles and fittings	BS EN 490
Slates	BS EN 12326-1
Felt	BS EN 13707

2.115 SAFETY CHAINS IN SEWERS

1. Mild steel safety chain shall be medium tolerance chain conforming to BS EN 818-3 Grade 4 nominal size 8 x 24. After manufacture, mild steel safety chains shall be hot dip galvanised in accordance with BS EN ISO 1461.

2. Stainless steel safety chain shall be manufactured from Grade X5CrNiMo 17-12-2 steel conforming to BS EN 10088-3. Chain links shall be welded and have an internal length not exceeding 45 mm and an internal width of between 12 mm and 18 mm. The fins caused by welding shall be removed and the weld shall be smoothly finished all round.

3. The chain shall have a breaking force of 30 kN and a proof force of 15 kN when tested in accordance with BS EN 818-1.

4. Anchor bolts for fixing safety chains shall be of the stainless steel safety type which provides a progressive mode of failure.

2.116 SAMPLE TAPS

(i) Taps should be specifically designed for sampling such as chrome-plated gunmetal with a tapered spout and protective cover, commonly referred to as the "Harris Tap".

(ii) Single non-rising spindle headworks tap to BS EN 200 with the flow straightener or aerator removed may suitable for this purpose.

1. Sample taps shall be metal complying with BS EN 200 or equivalent, without attachments or inserts.

2. For sample points located externally, the tap shall be enclosed in a lockable, purpose-designed sample pillar.

3. Taps for sampling purposes shall be a ½" male chrome sample tap, ½" BSP inlet. They shall be permanently labelled with "Sample Name/Water Quality Site No./Required Flushing Time".

2.117 SANDS

(i) The requirement for sands to be washed is additional to the requirements of the Standards but is in line with the main conclusion of CIRIA Report R59 'Building Sands: Availability, Usage and Compliance with Specification Requirements'.

1. Sands for mortar and grout shall comply with BS EN 13139 and PD 6682-3.

2. Sands for floor screeds shall comply with the relevant provisions of BS EN 12620 and PD 6682-1.

3. Sands for external rendering and internal plastering with lime and Portland cement shall comply with the relevant provisions of BS EN 13139 and PD 6682-3.

4. All sands required to comply with BS EN 12620 and PD 6682-1, BS EN 13139 and PD 6682-3 shall be washed sands.

2.118 SEWER LININGS

(i) Appendix A of each WIS, lists information to be established for particular design situations.

1. Sewer linings shall comply with the relevant provisions of the appropriate Standard, as set out in the following Table:

Type	WIS
Glassfibre reinforced cement (GRC)	4-12-04
Precast gunite	4-12-05
Precast and in-situ ferrocement	4-12-06
Polyethylene (PE) pipes (non-pressure applications)	4-32-05
Glassfibre reinforced plastics (GRP)	4-34-02
Cured-in-place pipe (CIPP)	4-34-04
Polyester resin concrete (PRC)	4-34-05

2.119 SOIL, WASTE AND VENTILATING PIPES

(i) BS EN 877 deals with flexible joints for cast iron soil, waste and ventilating pipes and fittings.

1. Soil, waste and ventilating pipes, fittings and accessories for above-ground drainage systems shall comply with the relevant provisions of the appropriate Standard, as set out below:

Material	Standard
Cast iron	BS 416-1 or BS EN 877
PVC-U (soil and ventilating)	BS 4514
Polypropylene (waste)	BS EN 1451-1
Plastics (waste)	BS EN 1329-1, BS EN 1455-1, BS EN 1519-1, BS EN 1565-1 and BS EN 1566-1

2. Wash-basin and sink wastes shall comply with the relevant provisions of BS EN 274-1.

2.120 STEEL REINFORCEMENT

1. Steel reinforcement shall comply with the relevant provisions of the appropriate Standard, as set out below:

Type	Standard
Carbon steel bars	BS EN 10080 and BS 4449
Steel wires	BS EN 10080 and BS 4482
Steel fabric	BS EN 10080 and BS 4483
Stainless steel	BS 6744
Epoxy coated steel	BS EN ISO 14654
Bed joint reinforcement for masonry	BS EN 845-3

2. Steel fabric reinforcement shall be welded at the intersections and shall be delivered in flat sheets, except where pre-bent reinforcement is specified.

3. Steel reinforcement shall be obtained from suppliers holding a valid Certificate of Approval for the manufacture and/or fabrication of steel reinforcement issued by the UK Certification Authority for Reinforcing Steels or equivalent authority. The CARES, or equivalent, Certificate of Approval Number shall be stated on all appropriate purchase documentation.

2.121 STEEL SHEET PILES

(i) The required grade of steel should be described in the Contract.

(ii) The requirement for piles to be coated should be deleted if no part is to be exposed on completion of the Works.

1. Steel from which steel sheet piles are rolled shall comply with the relevant provisions of BS EN 10025-1 and BS EN 10025-2.

2. Maximum rolling margins shall be 4% above and 2.5% below the calculated masses and 75 mm over and 50 mm under the required lengths.

3. Before being driven, permanent steel sheet piles shall be wire-brushed to remove loose rust and dirt and be coated with black tar-based paint complying with BS 1070, Type B, except that piles in contact with water to be used for potable supply shall be coated with black bitumen solution to BS 3416.

2.122 STILES, BRIDLE GATES AND KISSING GATES

1. Stiles, bridle gates and kissing gates shall comply with the relevant provisions of BS 5709.

2. All timber for stiles, bridle gates, kissing gates and posts shall be given preservative treatment in accordance with the provisions of BS 5709.

3. All fittings and steel stiles, bridle gates, kissing gates and posts shall be hot dip galvanised in accordance with BS EN ISO 1461.

4. Concrete for surrounding the base of posts shall be compressive strength class C16/20 (GEN3).

2.123 STRUCTURAL STEEL

1. Structural steel sections shall comply with the relevant provisions of the appropriate Standard, as set out in the following table:

Type	Standard
Hot-rolled structural steel sections	BS 4-1
Cold-rolled steel sections	BS EN 10162
Hot finished structural hollow steel sections	BS EN 10210-2
Structural steel angles	BS EN 10056-1

2.124 SURFACE BOXES AND GUARDS

(i) Particular requirements from the options listed in Appendix A of BS 5834-1 should be described in the Contract, as should those from Appendix B of Parts 2 and 3.

(ii) The minimum grade of cover should be described in the Contract to suit the vehicle intensity. BS EN 124 describes the minimum grades suitable for the location of manhole covers in carriageways, footways and other locations, e.g., minimum grade C250 when located within the kerb area (measured 0.2 m into the footway/verge and 0.5 m into the carriageway) and D400 when located within the carriageway area (measured 0.5 m from the kerb into the highway).

(iii) BS 6700 applies to meter boxes and their installation.

(iv) WIS 4-37-01 covers the specification of boundary boxes for the metering and control of domestic and small industrial water services.

1. Small and large surface boxes shall comply with the relevant provisions of BS 5834-2 and BS 5834-3, respectively, or WIS 4-37-01.

2. Small surface boxes having hinged lids shall be provided with hinge pins of a minimum diameter of 8 mm and made from stainless steel or copper alloy to prevent corrosion. Other means of retention of surface boxes shall be resistant to corrosion.

3. Guards and foundation units for underground stopvalves shall comply with the relevant provisions of BS 5834-1 or WIS 4-37-01.

4. Covers for the surface boxes on water mains shall have either the word "WATER" or the letter "W" cast on the top surface in 75 mm letters, as applicable. Covers for other applications shall similarly be marked: FH (fire hydrant), WO (washout hydrant), SV (sluice valve) and AV (air valve).

5. Covers and frames to be installed in carriageway or other areas with frequent passage of vehicles shall be to the minimum grade stated in BS EN 124 or Grade A to BS 5834-2.

6. Covers and frames to be installed in all other areas shall be a minimum Grade of B125 to BS EN 124, or Grade B to BS 5834-2.

7. Precast concrete sections for chambers and base units for buried waterworks apparatus up to and including 600 x 450 mm clear opening, shall comply with BS 5834-1 and BS 5834-4. All sections shall be Grade A as defined in that Standard.

8. Chambers of materials other than precast concrete shall meet the loading requirements in BS 5834-2.

9. In addition to the requirements of the above, domestic water meter boxes shall be of the waterproof type and shall contain an integral shut off valve on the upstream side together with non-return valve. They shall be suitable for use with concentric type flowmeters. They shall be adjustable for height and slope.

10. Surface boxes and guards providing direct access to areas of potable water shall be Loss Prevention Council (LPC) security rating 5 compliant.

2.125 SYNTHETIC RESIN ADHESIVES

1. Synthetic resin adhesives for plywood and wood shall be phenolic and aminoplastic, and comply with the relevant provisions of BS EN 301.

2.126 TIMBER AND PRESERVATION OF TIMBER

(i) See Regulations 8 and 10 of the Construction (General Provisions) Regulations 1961, as amended by the Construction (Metrication) Regulations 1984 for provisions relating to timber for excavations, shafts, tunnels and headings.

(ii) The use, class, service factor and desired life category should be described in the Contract.

1. All timber in the permanent Works shall be new. Timber for structural use shall be visually graded GS or SS to BS 4978 with an assigned strength class C16 to BS EN 1912 or machine graded as strength class C16 to BS EN 338.

2. Blockboards and other laminated woods shall comply with BS EN 12871.

3. Preservative treatment of timber shall be in accordance with BS 8417.

(iii) Consideration should be given to the use of reclaimed timber and any provision for its re-use should be described in the Contract.

(iv) The use of creosote as a preservative should be discouraged.

2.127 TREES AND SHRUBS

1. Trees and shrubs shall comply with the relevant provisions of the appropriate Standard, as set out below:

Type	Standard
Ordinary nursery stock	BS 3936-1
Semi-mature trees	BS 4043
Advanced nursery stock	BS 4043

2.128 TYING WIRE

1. Tying wire for steel reinforcement shall be 1.6 mm diameter, annealed mild steel wire complying with BS 1052.

2.129 UNPLASTICISED PVC PIPES AND FITTINGS

(i) The colour and size limitation for PVC pressure pipes are consistent with the current recommendations of the National Joint Utilities Group (NJUG).

(ii) WIS 4-31-08 deals with molecular oriented PVC-O pipes.

(iii) PVC pipe containing lead-based stabilisers are no longer acceptable under the Water Supply (Water Quality) Regulations.

1. PVC pressure pipes, joints and fittings shall comply with the relevant provisions, as set out below:

Material	Pipe	Joints and Fittings
PVC-U (metric blue)	BS EN ISO 1452-2	BS EN ISO 1452-3
PVC-A	PAS 27	PAS 27
PVC-O	WIS 4-31-08 (12.5 bar and 16 bar only)	BS EN ISO 1452-3

2. Pipes for potable water use shall be coloured blue.

3. PVC pipes, joints and fittings for gravity sewers and drains shall comply with the relevant provisions of BS EN 13598-1, BS 4660 or BS EN 1401-1.

4. Thermoplastics structured wall sewer pipe shall comply with the relevant provisions of BS EN 13476-1 and WIS 4-35-01 and BS EN 13476-2 or BS EN 13476-3 with the following properties:

- Maximum length of pipe for laying to be 3 m;
- Pipe Nominal Short Term Ring Stiffness to be not less than 8 kN/m^2;
- Long term Deformation to be less than 6% of the vertical nominal pipe diameter;
- Factor of safety against buckling to be not less than 2.5.

5. Solvent cements for jointing unplasticised PVC pipes shall comply with BS EN 14814. For pipes and fittings complying with BS 4660, BS EN 13598-1 or BS EN 1401-1, solvent cement may alternatively comply with BS EN 14680.

6. Solvent cements for jointing unplasticised PVC pipes shall not be used for below-ground use.

7. Push-fit joints shall be spigot and socket.

2.130 VALVES AND PENSTOCKS

(i) Particular requirements from the options listed in the various Standards should be described in the Contract.

(ii) BS 6683 deals with the installation and use of valves.

(iii) Clause 2.135 gives details of stopvalves.

(iv) BS EN 1092-1 and BS EN 1092-2 deal with flanges.

(v) The method used to control and operate valves and penstocks should be described in the Contract.

(vi) The responsibilities for supplying turn keys for operating those valves with caps should be detailed in the Contract.

(vii) The Contract should indicate the torque requirements for operating valves and penstocks.

(viii) Wedge or gate valves for water supply purposes should be selected with the operational and design features described in BS 5163-1 and BS 5163-2.

(ix) On water pipelines, it should be noted that all fittings are regarded as being in contact with potable water, see Clause 2.1.

1. Valves and penstocks for pipeline installation shall comply with the relevant provisions of the appropriate British Standard, as set out below:

Type	Standard
Isolating valves for water supply (includes wedge gate and butterfly)	BS EN 1074-1 and 2
Check/non-return valves for water supply	BS EN 1074-3
Air valves for water supply	BS EN 1074-4
Control valves for water supply	BS EN 1074-5
Industrial butterfly valves	BS EN 593
Penstocks	BS 7775
Cast iron industrial gate valves	BS EN 1171
Cast iron plug valves	BS 5158
Cast iron check valves	BS EN 12334
Diaphragm valves – metallic	BS EN 13397
Copper alloy globe, globe stop and check valves	BS 5154
Copper alloy gate valves	BS EN 12288

2. Valve and penstock parts to be in contact with potable water shall meet the relevant provisions, as set out below:

Type	Standard
Metallic	DD 256
Non-metallic	BS 6920-1

3. The surfaces of all valves and penstocks shall be protected from corrosion either by the nature of their material of construction or shall be coated in accordance with WIS 4-52-01. Internal water wetted surfaces shall be coated to Class A standard. All other surfaces shall be coated to Class B standard.

4. The direction of closure of all valves shall be agreed with the Water Undertaker and stated in the Contract. All valves shall have the direction of closing clearly indicated on the handwheel or body, as appropriate.

2.131 VITREOUS ENAMEL TANKS

(i) The coating type and thickness should be described in the Contract to suit the duty requirements.

1. Vitreous enamel (glass fused) steel tanks shall comply with WIS 4-25-01.

2. Vitreous enamel tanks shall be designed and the glass coating selected in accordance with BS EN 15282.

3. All components used in the construction of a tank shall have a design life of a minimum of 20 years for the conditions associated with the application, demonstrated in accordance with BS ISO 15686-2.

4. Each tank panel shall be tested to and conform with BS EN 15282.

5. The tank supplier shall provide certificates for each of the tank panels to confirm that all have been tested to the appropriate standard detailed in BS EN 15282 and to confirm that all panels supplied have passed the testing requirements.

6. Where it is necessary to construct pipework through the tank wall, the tank supplier shall provide galvanised steel, PN16 flanged connectors, complete with gaskets, internally and externally, which are compatible with the curvature of the tank. All cut-outs or sheet penetrations shall be formed in the raw steel sheet prior to the application of the vitreous enamel coating.

7. Where tanks are to be used for mixing of fluids, the tank shall be designed to withstand the dynamic loads from mixers and to resist abrasion from material within the tank contents (e.g., grit). This may take the form of a sacrificial baffle, or concrete baffle in front of the tank wall.

8. Where man entry into the tank is required at the external ground level, there shall be an 800 mm minimum clear opening. A bolted man entry hatch shall be provided unless otherwise stated in the Contract. The high strength low alloy (HSLA) steel hatch door shall be protected against corrosion to ensure a design life equivalent to the tank. All manways shall be supplied with davit arms.

2.132 VITRIFIED CLAY PIPES AND PIPELINE FITTINGS

(i) IGN 4-11-01 deals with vitrified clay pipes and fittings.

(ii) The type of joint and jointing materials for extra chemically-resistant pipes should be described in the Contract.

1. Vitrified clay pipes and pipeline fittings, including extra chemically-resistant pipes and fittings, shall comply with the relevant provisions of BS 65 or BS EN 295-1.

2. Vitrified clay jacking pipes shall conform to BS EN 295-7.

2.133 WALL TILES

(i) The type, size, thickness and colour of tiles should be described in the Contract.

1. Ceramic tiles for internal walls shall comply with the relevant provisions of BS EN 14411.

2.134 WATER

(i) In certain areas, supplementary mains carrying non-potable water have been laid. The use of this water with cement or in contact with potable water mains and installations has been prohibited.

(ii) Where potable water is not available, then alternative provision should be detailed in the Contract. BS EN 1008 gives details of suitability tests for water for use in making concrete.

1. Water for use with cementitious materials, or in contact with potable water mains and installations, shall be of potable quality.

2.135 WATER FITTINGS AND APPLIANCES

(i) On potable water pipelines it should be noted that all fittings are regarded as being in contact with potable water, see Clause 2.1.

1. Water fittings and appliances shall comply with the relevant provisions of the appropriate Standard, as set out in the following table:

(ii) Where possible, all fittings and appliances used as part of the supply of treated water for domestic purposes in premises should be approved by WRAS.

Water Fitting/ Appliance	Type/Material	Standard
Draining taps	Screw-down pattern	BS 2879
Draw-off taps	Metal bodied (performance)	BS EN 200
	Plastics bodied (performance)	BS EN 200
Stopvalves	Above-ground, screw-down pattern	BS 1010-2, BS 5433,
	Underground	WIS 4-23-04
Float operated valves	Diaphragm type (copper alloy body)	BS 1212-2
	Diaphragm type (plastics body)	BS 1212-3
Floats for ballvalves	Copper	BS 1968
	Plastics	BS 2456
Storage cisterns and lids	Low carbon steel	BS 417-2
	Thermoplastics	BS 4213
Wash basins	Ceramic	BS 1188
	Metal hand rinse	BS EN 111
Sinks	Glazed fireclay	BS 1206
	Stainless steel	BS EN 13310
WC pans	Horizontal outlet	BS EN 33, BS EN 37 and BS EN 997
Connectors for:		
WC pans	Plastics	BS 5627
WC seats and covers	Plastics, Type 1	BS 1254
WC flushing cisterns	Dual flush type	BS 1125
Urinals	Stainless steel slab	BS 4880-1
	Vitreous china bowl	BS 5520
Automatic flushing cisterns for urinals	Lidded	BS 1876
Ferrules	Various	WIS 4-22-02

2.136 WATERSTOPS

1. Rubber waterstops shall have the following properties:

Property	Test Standard	Requirements
Density	BS ISO 2781	1100 kg/m^3 (± 5%)
Hardness	BS ISO 48	60 - 70 IRHD
Tensile Strength	BS ISO 37	Not less than 17.5 N/mm^2
Elongation at fracture	BS ISO 37	Not less than 450%
Water absorption	BS ISO 1817	Not exceeding 5% (48 hours immersion)

2. Rubber waterstops shall be suitable for storage, handling, installation and service within a temperature range of 0°C to + 40°C.

3. Hydrophilic expanding waterstops shall have a delay coating to prevent premature swelling and shall have a minimum expansion of 1 to 3 volumetric change. They shall not be used in formed movement joints.

2.137 WET-MIX MACADAM

1. Wet-mix macadam shall consist of crushed rock or crushed slag, graded in accordance with the following table:

BS 410 Test Sieve	Percentage by Mass Passing
50 mm	100
37.5 mm	95 - 100
20 mm	60 - 80
10 mm	40 - 60
5 mm	25 - 40
2.36 mm	15 - 30
600 μm	8 - 22
75 μm	0 - 8

2. The particle size shall be determined by the washing and sieving method of BS EN 933-3.

3. Aggregate quality and cleanliness shall comply with the relevant requirements of Clause 2.4. The flakiness index shall be less than 35 when determined in accordance with BS EN 933-3.

4. The moisture content of the wet-mix macadam shall be the optimum ± 0.5%, as determined in accordance with BS 5835-1.

2.138 WINDOWS

1. Windows, window surrounds and fixings shall comply with the relevant provisions of the appropriate Standard, as set out in the following table:

Material/Type	Standard
Wood	BS 644
Steel	BS 6510
Aluminium	BS 4873
Plastics (PVC-U)	BS 7412

2.139 WINDOW SILLS

1. Precast concrete, cast stone, clayware, slate and natural stone window sills shall comply with the relevant provisions of BS 5642-1.

2. Steel window sills shall comply with the relevant provisions of BS 6510.

2.140 WOOD FLOORING

(i) The required type and finished thickness of flooring should be described in the Contract.

1. Tongued and grooved softwood board and strip floorings shall be graded and sized in accordance with the relevant provisions of BS 1297.

2.141 WOOD TRIM

(i) The relevant design reference in BS 1186-3 should be described in the Contract.

1. Wood trim in the form of architraves, skirtings, picture rails, cover fillets, quadrant, half-round and scotia moulds shall comply with the relevant provisions of BS 1186-3.

2.142 WROUGHT ALUMINIUM AND ALUMINIUM ALLOY

1. Wrought aluminium and aluminium alloys shall comply with the relevant provisions of the appropriate Standard, as set out below:

Type	Standard
Sections for structural purposes	BS 1161
Plate, sheet and strip	BS EN 485-1
Drawn rod, bar and tube	BS 754-1
Extruded rod, bar, tube and profiles	BS 755-1
Castings	BS 1559-1

SECTION 3

EXCAVATION, BACKFILLING AND RESTORATION

The Notes for Guidance are not part of the Specification.

3.1 EXCAVATION

1. Operations shall be carried out in such a manner as to prevent damage to, or deterioration of, the formation of excavations.

2. Excavation in roads and streets shall be carried out in accordance with the relevant Highway Reinstatement Specification.

3. Excavations in locations where services may be encountered shall be carried out in accordance with 'HSG 47 Avoiding Danger from Underground Services'.

4. Trenches shall be excavated so that the effective width is maintained within any limit imposed by the design of the pipeline. The sides of excavations shall be adequately supported at all times and, except where described in or permitted under the Contract, shall not be battered.

5. Unsuitable ground or damaged surfaces below formation shall be excavated and then filled to formation level, with the material shown on the drawings or otherwise specified in the Contract. Any void that results from over-excavation shall be filled with the material stated in the Contract.

6. Excavated granular material which can be reused shall be kept separate from excavated cohesive materials.

7. Trenches in rock for flexible pipes shall be excavated to provide a minimum clearance of 100 mm around the outside of pipe barrels and joints for pipes up to 100 mm nominal bore, and 150 mm for pipes of larger nominal bore. For rigid pipes the minimum clearance shall be 200 mm.

8. Trenches for pipes carrying water under pressure shall, except where otherwise described in the Contract, be excavated to a sufficient depth to ensure a minimum cover of 900 mm to the top of the pipes.

(i) The following publications give recommendations as to standards of good practice for excavation:

BS 6031;
BS 6164;
Report R97, 'Trenching Practice', published by CIRIA; and
Technical Note TN95 'Proprietary Trench Support systems', published by CIRIA.

(ii) A definition of rock shall be included in the Contract.

(iii) The relevant Highway Reinstatement Specification Roles should be defined in the Contract.

(iv) Any special requirements for Site clearance or for the disposal of excavated materials should be described in the Contract.

(v) The Contract should describe the extent of any excavations where battered sides will be permitted. A detail of the allowable cross-section should be given.

(vi) Excavation in carriageways should, wherever possible, be located such that the edge of the opening is at least 1 m from the edge of the carriageway.

(vii) Care should be taken when siting spoil heaps to avoid damaging trees by impinging on their root spread.

(viii) Excavated material for disposal off Site would normally be classified as controlled waste and the Client, under the Environmental Protection Act 1990, has a duty of care to ensure that the Works are carried out in compliance with the legislation relating to the treatment, keeping or disposal of such material.

(ix) Reference should be made to The Financing Act 1996 (Landfill Tax).

(x) Stored excavated granular material may require draining to achieve acceptable water content.

3.2 RELAYING TURF

1. Good practice dictates that turves shall be green when cut, they shall be kept moist and shall be re-laid in suitable weather conditions.

2. The level of topsoil beneath turves shall be such that the final grass surface after compaction shall be flush with the adjoining grass surface.

3.3 TOPSOIL FOR RE-USE

(i) It is advisable to make an assessment of soil stacking requirements in cases where topsoil quality is important, and to provide accordingly in the Contract

1. "Topsoil" shall mean the top layer of soil that can support vegetation. It shall include all turf not required for relaying or not acceptable for turfing under Clause 3.2.

2. Topsoil shall be removed from the areas described in the Contract and, where required for re-use, shall be stockpiled separately and kept free from weeds.

3. Handling of topsoil shall be in accordance with BS 3882.

4. Topsoil stripping, stockpiling and replacement shall be conducted in such a manner as to minimise damage to the soil structure.

3.4 DEALING WITH WATER

(i) Any requirements for prevention of the deposition of silt and/or protection from erosion should be set out in the Contract.

1. Water shall not be allowed to lie anywhere within excavations, unless so required under the Contract. Any drainage sumps required shall, where practicable, be sited outside the area excavated for the Works and shall be re-filled with approved material to the level of the underside of the adjacent permanent Works.

2. All necessary precautions shall be taken to prevent any adjacent ground from being adversely affected by loss of fines through any dewatering process.

3. Groundwater shall not be allowed to enter mains to be used for the conveyance of potable water.

4. The Contractor shall provide method statements and details of pollution prevention measures relating to the control and disposal of groundwater from de-watering operations. Discharging shall be subject to the Contractor obtaining prior written approval from the appropriate consenting body.

3.5 TEMPORARY DRAINS

1. Where temporary drains are required, they shall be laid in a narrow trench or grip formed below the bottom of the excavation in an approved position. The pipes shall be open-jointed and shall be surrounded with free-draining granular material.

2. When no longer required, temporary drains shall be removed or sealed.

3. When sealing temporary drains, grouting pipes shall be inserted in the line of the temporary drains at intervals not exceeding 25 m. The drains shall be solidly filled with grout (Class G3 or G4) and the grouting pipes cut off on completion of the filling. Care shall be taken to avoid impregnation of the granular bedding material around the main pipeline.

3.6 BACKFILLING

(i) Any special requirements for backfilling around mains and services should be described in the Contract.

(ii) Any particular requirements for the materials to be used for backfilling should be described in the Contract.

1. Backfilling shall, wherever practicable, be undertaken immediately the specified operations preceding it have been completed. Backfilling shall not, however, be commenced until the Works to be covered have achieved a strength sufficient to withstand all loading imposed thereon.

2. Backfilling shall be undertaken in such a manner as to avoid uneven loading or damage.

(iii) No materials should be stored on roofs of tanks without the written approval of the Client.

(iv) When considering the requirements for backfilling to water-retaining structures, reference should be made to Section 7.14 of this Specification.

3. Filling material to excavations not situated in the highway shall be in accordance with Clause 2.49, placed and compacted to form a stable backfill.

4. Excavations in roads and streets shall be filled above the level of any pipe surround required, in accordance with the relevant Highway Reinstatement Specification.

5. Where the excavations have been supported and the supports are to be removed these, where practicable, shall be withdrawn progressively as backfilling proceeds, in such a manner as to minimise the danger of collapse, and all voids formed behind the supports shall be carefully filled and compacted.

3.7 REINSTATEMENT OF MAINTAINABLE HIGHWAYS

(i) Any particular requirements for the reinstatement method, materials and depths of layers should be described in the Contract.

(ii) Reinstatement of surface boxes is covered in Clause 5.21.

(iii) Reference should be made to the New Roads and Street Works Act.

1. Reinstatement of roads and streets (including carriageways footways, footpaths, cycle tracks and verges) which are maintainable highways shall be undertaken in accordance with the relevant provisions of the relevant Highway Reinstatement Specification.

2. Road categories and reinstatement requirements shall be obtained from the relevant Highway Authority and comply with the New Roads and Street Works Act.

3. Kerbs, channels, edgings and quadrants disturbed by the Works shall be re-laid with existing units, providing they are not damaged. Where existing units are not suitable for re-use, replacement units of similar texture, colour and type, consistent with those adjacent and complying with the relevant provisions of Clause 2.35 or 2.98, as appropriate, shall be provided.

4. The re-laying of kerbs, channels, edgings and quadrants shall be in accordance with Clause 8.10. In-situ kerbs and channels shall be reinstated to conform with adjoining kerbs and channels.

3.8 REINSTATEMENT OF NON-MAINTAINABLE HIGHWAYS

(i) Where the relevant Highways Reinstatement Specification is inappropriate, the reinstatement should be described in the Contract.

(ii) Any discussions with the landowner should be carried out through a land agent.

1. Non-maintainable highways shall, except where otherwise stated in the Contract, be reinstated in accordance with the relevant provisions of Clause 3.7.

3.9 REINSTATEMENT OF UNPAVED LAND

(i) Any special grass seed mixtures required, differing from those specified in Clause 2.56, should be described in the Contract.

(ii) Any requirements to apply fertiliser should be described in the Contract.

(iii) Any special provisions for the reinstatement of land should be set out in the Contract, including the use of deep ploughing where overcompaction of the unpaved land has occurred.

1. Proposals for reinstatement shall be discussed with the landowner prior to any work commencing (see Section 1.7). Subsoiling of the working strip is required, unless otherwise stated in the Contract.

2. The surface of all land affected shall be broken up, to a depth of at least 300 mm, and stones and extraneous material greater than 50 mm in size cleared before topsoil is replaced. The land shall be cultivated and restored as closely as possible to its original condition.

3. Surfaces to be sown with grass seed shall be reduced to a fine tilth and cleared of stones and extraneous material greater than 50 mm in size. The seed shall be sown at the proper season, evenly distributed and applied at a rate of not less than the quantities given in the following table:

(iv) Any requirements for grass cutting and weed killing should be described in the Contract.

(v) Newly-turfed areas on slopes of cuttings and embankments may require to be secured.

Nature of Area to be Seeded	Level Surfaces (g/m^2)	Sloping Surfaces to Cuttings and Embankments (g/m^2)
Lawns	60	-
Surrounds to tanks and process plants	25	35
Agricultural land and roadside verges	6	10

4. Surfaces to be turfed shall be prepared as for seeding. Turves shall be placed, butted, interlocked and tamped, and the joints filled with fine sandy soil. On sloping ground, where they may be likely to slip, turves shall be laid diagonally. Any subsidence taking place shall be made good by taking up the turf, filling with good, finely-sieved topsoil and replacing the turf in the manner specified above. Any turf that dies shall be replaced with new turf.

5. Hydraulic mulch seeding shall be carried out using a proprietary process and shall contain types of seed, mulch material, fertiliser and other necessary additives to produce a covering of sward on subsoil.

3.10 TREES

(i) NJUG (National Joint Utilities Group) Publication Volume 4 'Guidelines for the Planning, Installation and Maintenance of Utility Services in Proximity to Trees', gives guidance on working near trees.

1. The planting, staking and maintenance of trees in the advanced nursery stock category shall be carried out in accordance with the relevant provisions of BS 4043.

2. The preparation, planting and securing of semi-mature trees shall be carried out in accordance with the relevant provisions of BS 4043 and their subsequent maintenance shall comply with BS 5837.

3. Tree surgery, repair work, bracing and feeding, and tree removal shall be carried out in accordance with the relevant provisions of BS 3998.

4. Measures to protect and preserve existing trees to be retained on the Site shall be taken in accordance with the relevant provisions of BS 5837.

3.11 REINSTATEMENT IN HIGHWAYS AND ROADS USING FOAMED CONCRETE

1. Reinstatement of openings in highways and roads using foamed concrete shall comply with the British Cement Association 'Foamed Concrete - A Specification for Use in Reinstatement of Openings in Highways'.

2. Reinstatement of openings in highways and roads using foamed concrete shall comply with:

(a) The 'Manual of Contract Documents for Highways Works – Volume 1 - Specification for Highway Works' Clause 1043; and

(b) The 'Specification for the Reinstatement of Openings in Highways' issued by the Highways Authorities and Utilities Committee.

3. The pipe surround material shall be protected from the foam concrete by an impermeable layer.

3.12 LAND DRAINS

(i) Any special requirements necessary to facilitate the restoration of land drainage should be described in the Contract.

1. The positions of all land drains intercepted, disturbed or reinstated shall be prominently marked at every point of intersection with the work. Records shall be kept of positions, depths, pipe diameters and the types of construction, and a copy of these records shall be given to the Client. Care shall be taken to prevent the disturbance of markers.

2. Prior to the permanent reinstatement of land drainage, existing drains, where intercepted by excavations, shall be cleared. Facilities shall be afforded to the Client and the landowner or occupier to inspect them and to determine the extent of replacement that may be necessary.

3. The backfill of intercepting excavations shall be compacted in 200 mm layers, to give a firm bearing immediately before replacement pipes are laid, and shall be brought up to the level of the underside of the land drains or of any support to be provided.

4. The affected land drains shall be cut back into firm ground until, in each case, a section is exposed which is unaffected by the Works.

5. Replacement pipes or support beams shall bear on undisturbed ground for at least 500 mm at each end. The replacement pipes shall be of the same internal diameter as the sections of drain which they replace and shall be properly connected at each end.

3.13 FILLING ABOVE GROUND

(i) This is a general Clause to cover filling which performs no specific load-bearing or structural role. A more thorough specification may be necessary in other cases (see, for instance, the Highways Agency's 'Specification for Highway Works').

1. Embankments and other areas of fill shall be formed of suitable materials capable of normal compaction to form a stable fill, deposited and compacted as soon as practicable after excavation, in layers of thickness appropriate to the compaction plant used.

2. The filling shall, where practicable, be built up and compacted evenly, and shall be maintained at all times with a sufficient camber or cross fall and a surface sufficiently even to enable surface water to drain readily from it.

3.14 BLASTING

(i) It may be necessary for the Contract to describe safe values for vibrational amplitude and peak particle velocity.

(ii) For further guidance on blasting with respect to tunnelling works, see British Tunnelling Society and Institution of Civil Engineers 'Specification for Tunnelling', Clause 309.

1. The procedure for using explosives shall be stated in the Contract.

2. Where blasting is proposed adjacent to a building or other structure, existing or under construction, the agreed safe values of vibrational amplitude and peak particle velocity shall not be exceeded.

3. The use of electrical detonators in the vicinity of static and mobile radio transmitters, including normal radio and television broadcasting stations and radar units associated with aircraft movements, shall comply with the provisions of BS 6657.

4. The handling, transport and use of explosives shall be in accordance with the relevant provisions of BS 5607. Explosives shall be used in the quantities and manner recommended by the manufacturer.

3.15 PILING

(i) The ICE 'Specification for Piling and Embedded Retaining Walls' includes 'Associated Guidance on Contract Documentation and Measurement'.

1. The Specification for piling shall be the 'Specification for Piling and Embedded Retaining Walls', published by the Institution of Civil Engineers in 1996.

3.16 DEMOLITION

(i) Any further information required should be stated in the Contract.

(ii) Reference should be made to The Health and Safety Executive guidance on demolition works (www.hse.gov.uk).

1. Demolition shall be carried out in accordance with BS 6187.

SECTION 4

CONCRETING AND FORMWORK

The Notes for Guidance are not part of the Specification.

4.1 SUPPLY OF INFORMATION

1. Before any concrete is supplied and not less than 7 days before the start of the concrete production, all pertinent information specified in BS EN 206-1:2000 Clause 7.2 and BS 8500-2:2006 Clause 5.2 shall be exchanged and agreed with the Producer.

4.2 INITIAL TESTING

(i) Where it is not practicable to carry out full-scale trials, special reference should be made in the Contract to laboratory-scale mixes.

(ii) Sufficient information should be derived from the initial testing to ensure that the concrete will meet the specified requirements. It may also be necessary to specify water absorption tests for structures designed to retain an aqueous liquid.

1. Initial testing shall be undertaken for each classification of structural concrete.

2. Such testing shall be in accordance with BS EN 206-1:2000 Clause 9.5, Clause 10 and Annex A.

4.3 IDENTITY TESTING

1. Where identity testing is specified for the testing of slump, flow and air content of individual batches, it shall be undertaken in accordance with BS 8500-1.

2. Where specified, identity testing for strength shall be carried out in accordance with Clause 2.20.5.

4.4 POROUS NO-FINES CONCRETE

1. The concrete shall not be mechanically vibrated or excessively worked when placed.

4.5 TRANSPORTING, PLACING AND COMPACTING

(i) Any requirements for placing concrete in special sequence, e.g., by alternate bay construction, should be described in the Contract.

(ii) "Adequate" in the context of notice to the Client should be sufficient time to allow the Client to fulfil their quality management obligations. The minimum time period should be agreed at the commencement of the Contract.

(iii) Any variations to the times in Clause 4.5.3 should be specified in the Contract.

(iv) Any requirements for testing fresh concrete should be specified as in Clause 2.20.5.

1. Concrete shall be transported from the mixer in accordance with BS 8500-2 and placed in the Works as rapidly as practicable, by methods which will prevent the segregation or loss of any of the ingredients and will maintain the required consistency. It shall be deposited, as nearly as practicable, in its final position and all equipment for transporting concrete shall be kept clean.

2. Adequate notice shall be given to the Client of the intention to commence concreting.

3. Concrete shall be delivered to Site within the times specified in Clause 14.2 of BS 8500-2: 2006.

4. Concrete shall be thoroughly compacted in its final position within 30 minutes of commencing discharge. The plant used for compaction shall be operated continuously during the placing of each batch of concrete until the expulsion of air has virtually ceased, and in a manner which does not promote segregation of the ingredients.

5. Whenever vibration has to be applied externally, the design of formwork and disposition of vibrators shall be such as to ensure efficient compaction and to avoid surface blemishes.

4.6 CONCRETING IN COLD WEATHER

(i) For further information see 'Winter Concreting', published by Concrete Information Ltd in 1985.

1. Concreting at ambient temperatures below 2°C may be carried out only if the following conditions are met:

(a) the aggregates and water used in the mix shall be free from snow, ice and frost;

(b) before placing concrete, the formwork, reinforcement and any surface with which the fresh concrete will be in contact shall be free from snow, ice and frost and shall be at a temperature above 0°C;

(c) the initial temperature of the concrete at the time of placing shall be at least 5°C as defined in BS EN 206-1:2000, Section 5.2.8 Lower Limit;

(d) the temperature at the surface of the concrete shall be maintained at not less than 5°C at any point until the concrete reaches a strength of 5 N/mm^2, as confirmed by tests on cubes matured under similar conditions; and

(e) temperatures at the surface of the concrete shall be measured where the lowest temperature is expected.

2. Precautions shall be taken to prevent the temperature of any concrete falling to 0°C during the first five days after placing.

4.7 CONCRETE TEMPERATURE

(i) The Contract should specify a reduced upper temperature if the design of the structure requires a lower temperature, i.e., to minimise thermal cracking on large pours.

1. The resultant temperature of the combined materials in any batch of concrete at the point and time of delivery to the Works shall not exceed the upper limit as stated in BS 8500-2. Cement shall not be permitted to come into contact with water at a temperature greater than 60°C.

2. Where the temperature of the fresh concrete is likely to exceed that specified in Clause 4.7.1, concreting shall not be permitted unless stated in the Contract.

4.8 CURING

(i) Consideration may have to be given to measures to prevent thermal cracking where a temperature differential in excess of 20°C is likely to occur, e.g., by extending striking times for the formwork.

1. Curing of concrete shall be carried out in accordance with BS EN 13670 Curing Class 2.

2. In cold weather, when the temperature of freshly-placed concrete may approach 0°C, cold water curing shall not be employed.

3. Components which are intended to have a similar exposed surface finish shall be cured in the same manner.

4.9 RECORDS OF CONCRETING

1. Up-to-date records of the dates and times when concreting is carried out, and of the weather and temperatures at those times, shall be kept. The records shall be available for inspection.

2. Sufficient records shall be maintained to enable every batch of concrete and its location within the Works to be identified.

4.10 CONSTRUCTION OF FORMWORK

(i) BS 5975 gives recommendations as to standards of good practice in formwork construction.

(ii) The positioning and detailing of movement joints should be described in the Contract.

(iii) Any special conditions relating to the re-use of forms, insofar as the materials of construction and repairs between uses may affect the colour and surface finish of exposed surfaces, should be

1. Formwork shall be sufficiently rigid and tight to prevent loss of mortar from the concrete and to maintain the correct position, shape and dimensions of the finished work. It shall be so constructed as to be removable from the cast concrete without shock or damage.

2. The forms shall be capable of producing a consistent quality of surface, as described in the Contract.

3. Where holes are required in forms to accommodate projecting reinforcement, fixing devices or other built-in items, precautions shall be taken to prevent loss of the mortar matrix.

4. Formwork shall give access for the preparation of joint surfaces before

described in the Contract.

(iv) Any special requirements regarding chamfers to internal angles should be described in the Contract.

the concrete has hardened.

5. For the purposes of compliance with the provisions of Clause 4.12.3, the method of constructing formwork shall allow for props to soffit forms to remain in position continuously for the period described.

6. All exposed vertical and horizontal edges of concrete shall have 25 mm x 25 mm chamfers. Chamfers shall extend to 150 mm below finished ground level.

4.11 CLEANING AND TREATMENT OF FORMS

1. The interiors of all forms shall be thoroughly cleaned out before any concrete is placed. The faces of the forms in contact with the concrete shall be clean and treated with a suitable release agent, where applicable.

2. Where a concrete surface is to be permanently exposed, only one release agent shall be used throughout the entire area. Release agents shall be applied evenly and contact with reinforcement and other embedded items avoided. Where the concrete surface is to receive an applied finish, care shall be taken to ensure the compatability of the release agent with the finish.

4.12 STRIKING OF FORMWORK

(i) As the removal of formwork is dependent upon the method of working, the Contract may state a formal procedure for determining striking times, based on CIRIA Report R136.

(ii) Interpretation of CIRIA Report R136 should be in accordance with the following:

(a) for members with "Rough finish", the requirements applicable to F1 and F2 finish apply.

(b) for members with "Fair finish" or "Fair worked finish", the requirements applicable to F3 and F4 finish apply.

(iii) "Rough finish", "Fair finish" and Fair worked finish" are described in Clause 4.22 (i).

1. Formwork shall be removed without shock to, or disturbance of, the concrete.

2. Formwork to vertical surfaces or sloping formwork not supporting concrete in flexure shall not be removed until the concrete strength shall be sufficient to meet any wind loading upon the concrete likely to arise at the time when the formwork is removed, and:

(a) the concrete strength (as confirmed by tests in cubes cured under representative conditions) has reached 5 N/mm^2; or

(b) for concrete containing cement to BS EN 197-1:2000 CEM I 42.5, 52.5 only, in the absence of cube test results, a minimum period shall have elapsed since the concrete was poured equivalent to 11 hours at 15°C for unsealed plywood forms or 8 hours at 15°C for impermeable forms.

3. Formwork supporting concrete in flexure shall not be removed until:

(a) the concrete strength (as confirmed by tests on cubes cured under representative conditions) has reached 10 N/mm^2, or twice the stress to which the concrete will then be subjected, whichever is the greater; or

(b) for concretes containing cement to BS EN 197-1:2000 CEM I 42.5, 52.5, in the absence of cube test results or any formal procedure agreed in writing, the periods before striking calculated from the relevant formula given in the following table shall be used.

Type of Formwork	Period Calculated for Mean Ambient Temperature (t) Between 0°C and 25°C Using Formulae Below
Soffit forms to slabs and beams	$\frac{100}{t + 10}$ days
Props to slabs and beams	$\frac{250}{t + 10}$ days

(iv) It should be noted that 11 hours at 15°C is equivalent to:
8 hours at 20°C
15 hours at 10°C
24 hours at 5°C

8 hours at 15°C is equivalent to:
6 hours at 20°C
12 hours at 10°C
18 hours at 5°C

4. For concrete containing cement not conforming to BS EN 197, the times for striking of formwork shall be derived from CIRIA Report R136.

5. Sufficient records to identify the time from pouring of any section to the striking of the formwork on the same shall be maintained on Site for inspection.

(v) Any requirement for the control of thermal cracking should be described in the Contract.

4.13 SLOPING FORMWORK

(i) Any requirement for top formwork at slopes flatter than 30° to the horizontal should be described in the Contract.

1. Top formwork shall be provided to slopes 30° or more from the horizontal, except where placing and compaction can achieve the specified surface finish without top formwork.

4.14 CUTTING AND BENDING OF REINFORCEMENT

(i) The Contract should detail any particular requirements for dealing with health and safety issues in connection with starter bars and the fixing of reinforcement, as a result of any designer's risk assessment.

1. Cutting and bending of reinforcement shall be in accordance with BS 8666.

2. Except for the use of proprietary starter bar systems, reinforcement shall not be straightened or re-bent. When using proprietary systems and bending projecting reinforcement, care shall be taken not to damage the concrete and to ensure that the radius is not less than the minimum specified in BS 8666.

4.15 FIXING OF REINFORCEMENT

(i) Where the use of proprietary reinforcement continuity systems, rolled mats, mesh or couplers is proposed as an alternative to the reinforcement detailed on the Contract Drawings, this could require a variation from the Contract.

1. Reinforcement shall be firmly supported and secured against displacement, in accordance with BS 7973-2.

2. Non-structural connections for the positioning of reinforcement shall be made with tying wire or other fixing devices. Projecting ends of ties or clips shall not encroach into the concrete cover.

3. Concrete cover shall not be less than the nominal cover minus 10 mm or greater than the nominal cover plus 15 mm (Δc in BS 8500-1:2006 A3). Unless shown otherwise on the drawings, nominal cover shall be:

(a) 45 mm for tops of walls;

(b) 50 mm for concrete cast against blinding;

(c) 75 mm for concrete cast against soil; and

(d) 40 mm for all other locations.

4.16 SURFACE CONDITION OF REINFORCEMENT

(i) Any protection required for steel left projecting should be described in the Contract.

1. Concrete shall not be placed until reinforcement is free from any substance which might adversely affect the steel or concrete chemically, or reduce the bond.

4.17 LAPS AND JOINTS

1. Laps and joints in reinforcement shall be made only at the positions described on the Drawings.

4.18 WELDING OF REINFORCEMENT

1. Reinforcement shall not be welded on site except where described in, or permitted under, the Contract. All welding procedures shall be stated in the Contract.

4.19 BUILT-IN ITEMS

(i) Some coatings may not provide good adherence to concrete and, therefore, may require additional sealing systems and/or features to ensure a sound mechanical key to the concrete structure The fittings should be designed to avoid formation of air pockets or voids during the placing/compaction of concrete.

1. Where pipes, sleeves, water bars or other items are designed to be cast into concrete, they shall be constructed and installed in order to ensure that they are:

(a) securely and rigidly retained in position;

(b) watertight for the design life of the fittings and structure;

(c) resistant to corrosion for the design life;

(d) adequately bonded to the concrete; and

(e) free from external coatings that may reduce the bond.

2. Precautions shall be taken to prevent the formation of air pockets, voids or other defects whilst the concrete is being placed.

4.20 CONSTRUCTION JOINTS

(i) Guidelines on the control of cracking can be found in CIRIA C660 'Early-age Thermal Crack Control in Concrete'.

(ii) Proprietary systems may be available for this application.

1. Position and details of construction joints in concrete shall be described in the Contract.

2. Joint lines shall be arranged to coincide, wherever possible, with features of the finished work.

3. Concreting shall be carried out continuously up to construction joints.

4. Concrete shall not be allowed to taper off to a thickness of less than 50 mm. Vertical joints shall be formed against a stop board, suitably notched to accommodate the reinforcement. The top surface of each lift of concrete shall be straight and level.

5. Where a kicker is used, it shall be at least 70 mm high and shall be cast integrally with the slab.

6. The surface of any concrete against which new concrete is to be cast shall be free from laitance and shall be roughened to the extent that the large aggregate is exposed but not disturbed. The joint surface shall be cleaned immediately before the fresh concrete is placed against it.

7. Where practicable, such preparation of joints shall be carried out when the concrete has set but not hardened.

4.21 SURFACE FINISHES PRODUCED WITHOUT FORMWORK

(i) The four finishes are intended to be applied as follows:

(a) Screeded - surfaces to receive further treatment or of no visual merit, or expressly suitable to their function with the workmanship as specified;

(b) Wood Float - surfaces where a reasonably regular finish is required but appearance is not of prime importance;

(c) Steel Trowel - surfaces where appearance is important; and

(d) Power Float Finish - surfaces where a smooth flat finish with only minor deviations is required.

(ii) Any other required finish should be described in the Contract.

1. The concrete shall be levelled and screeded to produce a uniform plain or ridged surface, as required. No further work shall be applied to the surface unless it is a first stage for a wood float or steel trowel finish.

2. The screeded finish shall be wood floated under light pressure to eliminate surface irregularities.

3. When the moisture film has disappeared and the concrete has hardened sufficiently to prevent laitance from being worked to the surface, the surface to the wood float finish shall be steel-trowelled under firm pressure to produce a dense, smooth, uniform surface free from trowel marks.

4. Power floating shall be undertaken by steel floating the concrete to an even finish with no ridges or steps. When the concrete has taken a primary set, it shall be power trowelled to a uniform smooth polished surface free from trowel marks or other blemishes. Once power floating is completed, the surface finish must be adequately protected from construction traffic.

5. Where the type of finish is not given, it shall be wood float finish.

4.22 SURFACE FINISHES PRODUCED WITH FORMWORK

(i) The three types of surface finish are intended to be applied as follows:

(a) Rough - surfaces next to earth, or to receive further treatment, or of no visual merit, or expressly suitable to their function with workmanship, as specified;

(b) Fair - surfaces required for serviceability and structural soundness and which are not visually important; and

(c) Fair worked - aqueous liquid retaining faces and other surfaces to good quality concrete, required for serviceability, structural soundness and appearance.

1. Rough finish shall be obtained by the use of moulds or properly designed forms of closely-jointed sawn boards. The surface shall be free from substantial voids, honeycombing or other large blemishes.

2. Fair finish shall be obtained from forms designed to produce a hard smooth surface with true, clean arrises. Only very minor surface blemishes shall be permitted and there shall be no staining or discolouration. Any projections shall be removed and the surface made good.

3. Fair worked finish shall be obtained by first producing a fair finish and then filling all surface blemishes with a fresh, specially prepared cement and fine aggregate paste whilst the concrete is still green, where possible. After the concrete has been properly cured, the faces shall be rubbed down, if required, to produce a smooth and even surface. If the surface is to be exposed in the final work, every effort shall be made to match the colour of the concrete.

(ii) If test panels are required, these should be described in the Contract.

(iii) CIRIA report C511 gives guidance on the type of CPF liners and their use.

4. Liquid retaining surfaces and other surfaces exposed in the completed Works shall receive a fair worked finish. All other structural concrete surfaces shall receive a fair finish.

5. The use of controlled permeability formwork (CPF) shall be agreed with the Client. The CPF liner shall have the following minimum improvements to the outer 20 mm of the concrete:

a) cement content of 75 kg/m^3 greater than the control; and

b) a minimum improvement of 50% for surface tensile strength, ISAT carbonation and chloride ingress.

6. The CPF liner shall be installed and used in accordance with the manufacturer's recommendations. Release agents shall not be used. Type III, CPF liners can be reused, subject to agreement with the Client.

4.23 WEARING SCREEDS

(i) The class of abrasion resistance required should be specified in the Contract.

(ii) Reference should be made to BS 8500-2.

1. Wearing screeds (granolithic finish) shall be provided, laid and finished in accordance with the relevant provisions of BS 8204-2.

2. Wearing screeds shall provide abrasion resistance to class AR4/WS of Table 4 of BS 8204-2:2003. Where high abrasion conditions are expected, wearing screeds shall provide abrasion resistance to class AR1/WS of Table 4 of BS 8204-2:2003.

3. Where concrete benching is required to have a granolithic finish with abrasion resistance to class AR4/WS of Table 4 of BS 8204-2:2003, this shall be formed with four parts 8 mm to dust to one part sulphate resisting cement placed with a steel trowel finish. Where sulphate resisting cement is unavailable, then a combination of Portland Cement (CEM 1) and GGB FS or PFA shall be used to give equivalent resistance to sulphate attacks.

4.24 TIE BOLTS FOR FORMWORK

1. Tie bolts shall be of the high tensile variety, fixed perpendicular to the formwork.

2. Tie bolts shall not embed any permanent metal parts within 50 mm of the concrete surface.

3. Voids remaining after the removal of all, or part, of each tie bolt shall be sealed using a polymer-modified cementitious compound, or other suitable product, unless specified in the Contract. Metal-based expansive admixtures shall not be used. All such voids shall be prepared in accordance with the manufacturer's instructions, prior to filling, to ensure an adequate bond is achieved.

4. Tie bolts which form a continuous hole through a structure designed to retain an aqueous liquid shall not be used.

5. In the case of structures designed to retain an aqueous liquid, any other measures securing formwork shall not impair the watertightness of the structure.

4.25 TOLERANCE FOR CONCRETE SURFACES

(i) The table applies to general concrete structures. Where more stringent tolerances are required, these should be described in the Contract (e.g., measuring flumes or areas where plant is to be installed).

1. Concrete surfaces in the final work shall have no abrupt irregularities which are, to an extent, observable by eye. Subject to retaining the required concrete cover to reinforcement, other deviations from the surfaces described in the Contract shall be no more than the following permissible amounts:

Type of Finish	Deviation From Line, Level, Vertically, Cross-sectional Dimension or Length (mm)
Screeded or rough	10
Power float	3
Any other	5

4.26 GROUT QUALITY CONTROL TESTING

(i) Grout strength requirements should be described in the Contract.

(ii) Grout can alternatively be specified by nominal mixes (see Clause 2.17).

1. Where tests are required for different properties of grout, they shall be carried out on samples from the same batch.

2. The density and workability of every batch shall be determined. The density shall not differ from the value described in the Contract by more than 5%. The workability shall not differ by more than 125 mm for the "Concrete Flow Through Test" or 5 seconds for the "Marsh Cone Test" with 10 mm orifice, from the values described in the Contract.

3. Sampling shall be at the rate of three cubes taken from every 5 m^3 of grout or 50 m of grouted annulus, whichever is smaller. When tested in accordance with the relevant provisions of BS 4551, cubes shall have a compressive strength, as given in the following table:

Grout Function	Minimum Compressive Strength at 28 Days (N/mm^2)
Annulus filling:	
Type I lining	12
Type II lining	3
Exterior void filling	2

4. Cube moulds shall be 70 mm (nominal) or 100 mm, and all joints shall be sealed to prevent leakage.

5. Moulds shall be overfilled and air bubbles removed by lightly tapping the mould. After leaving for 30-60 minutes, the excess grout shall be struck off and the moulds covered with plastic sheeting or damp hessian. Moulds shall be stored at 20°C ± 5°C for 24 hours or until the grout has attained sufficient strength to allow the cube to be stripped from the mould, whichever is the greater.

6. The cubes shall be removed from the moulds, marked, and stored in water at a temperature of 20°C ± 1°C until tested.

7. Trial mixes of the grout type proposed shall be undertaken to establish the properties of the grout. The amount of bleedwater shall be determined by filling a 100 mm diameter impermeable pot to a depth of 100 mm, covered to prevent evaporation. The bleedwater shall be removed after 3 hours. The percentage volume of bleedwater after 3 hours (expressed as volume of bleedwater:volume of original sample) shall not exceed:

(a) 1% for Type I linings; or

(b) 5% for Type II linings.

4.27 FIBRE REINFORCED CONCRETE (FRC)

(i) Polypropylene or stainless steel fibres may be used.

1. Where concrete reinforced with synthetic polymer or stainless steel fibres is used as an alternative to air-entrained concrete containing normal fabric reinforcement in roads and footpaths, this shall be agreed with the Client prior to commencement of the design.

2. Fibre reinforced concrete (FRC) shall be designed in accordance with Concrete Society Technical Report 63, 'Guidance for the Design of Steel-Fibre Reinforced Concrete'.

4.28 CONCRETE REPAIRS

1. All concrete repairs shall be undertaken in accordance with BS EN 1504-10. Test sections shall be required in order to determine the sufficiency and compatibility or otherwise, of the proposed materials and methods.

SECTION 5

CONSTRUCTION OF PIPELINES AND ANCILLARY WORKS

The Notes for Guidance are not part of the Specification.

5.1 PIPELAYING GENERALLY

(i) The following Publications give recommendations on standards of good practice for the installation of pipelines on land:

General: BS EN 14161

Ductile iron: BS 8010 Section 2.1

GRP: BS 8010 Section 2.5

Drains and sewers: BS EN 752.

(ii) Further information on installing box culverts is available in 'Guide to Site Practice', published by the Box Culvert Association.

(iii) Marker tape is not applicable to trenchless pipe installation.

(iv) The minimum depth of cover should be specified in the Contract.

1. Where socketed pipes are required to be laid on a granular or sand bed, or directly on a trench bottom, joint holes shall be formed in the bedding material or excavated formation to ensure that each pipe is uniformly supported throughout the length of its barrel and to enable the joint to be made.

2. Pipes shall be laid on setting blocks only where a concrete bed or cradle is used.

3. Where pipes are required to be bedded directly on the trench bottom, the formation shall be trimmed and levelled to provide even bedding of the pipeline and shall be free from all extraneous matter that may damage the pipe, pipe coating or sleeving.

4. Any protective cap, disk or other appliance on the end of a pipe or fitting shall only be removed permanently when the pipe or fitting which it protects is about to be jointed. Any exposed pipe ends shall be capped when pipelaying is not actively being carried out to prevent vermin or soil entering the pipework.

5. Pipes and fittings, including any lining or sheathing, shall be examined for damage and the joint surfaces and components shall be cleaned immediately before laying. Where repairs are required, these shall be agreed with the Client.

6. Suitable measures shall be taken to prevent extraneous material from entering the pipe, and to anchor each pipe to prevent flotation or other movement before the Works are complete.

7. Pipeline marker tape, in accordance with Clause 2.76, shall be laid between 100 mm and 300 mm above the pipe. Where a tracer system is specified, it shall be continuous and adequately secured to valves and fittings.

8. Where the gradient of the as-laid pipeline exceeds 5%, installation shall be uphill with sockets leading.

5.2 PIPE BEDDING

(i) When puddled clay stanks are required, these should be described in the Contract.

(ii) Details of pipe bedding, surround and sidefill should be described in the Contract.

(iii) Refer to BS EN 1295-1 for pipe bedding design details.

1. Bedding for pipes shall be constructed by spreading and compacting granular bedding material over the full width of the pipe trench. After the pipes have been laid, additional material shall, if required, be placed and compacted equally on each side of the pipes and, where practicable, this shall be done in sequence with the removal of the trench supports.

2. Where specified in the Contract, stanks shall be constructed in the pipe trench to inhibit the flow of groundwater along the granular bedding and pipe surround. Stanks shall be formed around pipelines on either side of stream crossings.

3. The stanks shall be constructed of either clay, which shall be puddled, placed, and compacted around the pipe, or a lean mix concrete which shall be placed and compacted around the pipe, over a length specified in the Contract. The requirement in both cases is that the stank shall surround the pipe to the bottom and to the full width of the excavated trench to form a barrier within the granular bed and surround and protrude into the backfill material above by an amount specified in the Contract.

4. In bad ground conditions where the migration of the pipe granular surround into the ground may occur, the surround shall be wrapped in geotextile membrane.

5.3 CONCRETE PROTECTION TO PIPES

(i) Rapid hardening cement should not be used in concrete for the protection of plastics pipes.

(ii) Plastics pipes should be wrapped with a layer of plastic sheeting complying with Clause 2.95 before being surrounded by concrete.

1. Pipes to be bedded on, or cradled with, concrete shall be supported on precast concrete setting blocks, the top face of each block being covered with two layers of compressible packing, in accordance with Clause 2.19.

2. Concrete provided as a protection to pipes shall not be less than Grade GEN 3, placed to the required depth in one operation.

3. Where pipes with flexible joints are used, concrete protection shall be interrupted over its full cross-section at each pipe joint by a shaped compressible filler, in accordance with Clause 2.19.

4. Where pipes are protected by a concrete slab placed above the pipe, this shall span the pipe trench and extend a minimum of 300 mm both sides, widening the trench above the pipe surround. There shall be a minimum of 150 mm of surround between the crown of the pipe and underside of the slab.

5.4 PIPE SURROUND

(i) Further guidance is available in BS EN 1295-1.

(ii) The minimum side gaps for narrow trenching techniques are stated for compaction purposes.

1. Pipe surround material shall, where required, be placed and compacted over the full width of the trench in layers not exceeding 150 mm before compaction, to a finished thickness of between 100 mm to 300 mm above the crown of the pipes.

2. Subsequent filling shall then be carried out, as specified in Clause 3.6.

3. Where trenching excavators or similar narrow trenching techniques are employed for open dig laying of pipelines, the minimum gap between the pipe barrel and side of the trench shall be 30 mm for pipe diameters of 280 mm or less, and 50 mm for pipes greater than 280 mm diameter.

5.5 PIPELAYING IN HEADINGS

(i) See also Clauses 3.1, 5.1 and 5.3.

1. Pipes to be laid in headings shall be supplied in lengths suitable for handling, jointing and packing within the working space available.

2. Headings shall be driven from shaft to shaft, or in such other lengths as may be described in the Contract, before any pipelaying is commenced.

3. After pipelaying, headings shall be packed solid with dry lean mix concrete so as to fill all voids. Where manual packing is employed, each pipe shall be surrounded before laying and jointing the next pipe.

4. Where grouting of headings is described in the Contract, grouting pipes shall be left in the top of the heading projecting behind each head tree and the whole grouted solid with grout Class G1. Grouting shall be carried out at the end of each shift or after three settings have been packed, whichever is the shorter interval.

5.6 THRUST BLOCKS

(i) Thrust blocks should be described in the Contract.

(ii) Refer to CIRIA Report TN 128 for the construction and design of thrust blocks.

1. Except where welded steel pipelines, welded polyethylene pipelines or self anchoring joints are used, thrusts from bends and branches in pressure pipelines shall be resisted by concrete thrust blocks cast in contact with undisturbed ground.

2. Any additional excavation required to accommodate thrust blocks shall be carried out after the bend or branch is in position and the thrust face shall be trimmed back to remove all loose or weathered material immediately prior to concreting.

3. Thrust blocks shall be allowed to develop adequate strength before any internal pressure is applied to the pipeline.

4. Rapid hardening cement shall not be used in concrete for thrust blocks to plastics pipes.

5. Plastics pipes shall be wrapped with a layer of plastic sheeting complying with Clause 2.95 before being surrounded by concrete.

6. Where connections to existing mains incorporate the casting of thrust or other support blocks, the temporary support shall be provided to pipework until such time as the concrete has cured sufficiently to resist the effects of forseeable pressure in the affected mains.

7. The depth of cover to concrete blocks shall not be less than 600 mm, unless otherwise stated in the Contract.

8. Where a concrete surround to the pipe barrel is to be provided as anchorage, a means shall be provided to transmit the load to the concrete, e.g., via an integral stub flange or similar fitting to form a flange.

5.7 PIPE JOINTING GENERALLY

(i) Any special requirements for filling the joint annulus should be described in the Contract.

(ii) The remaining flexibility is required for any subsequent settlement or ground movement.

1. Pipe jointing surfaces and components shall be kept clean and free from extraneous matter until the joints have been made or assembled. Care shall be taken to ensure that there is no ingress of grout or other extraneous material into the joint annulus after the joint has been made.

2. Where pipes with flexible joints are required to be laid to curves, the deflection at any joint as-laid shall not exceed three quarters of the maximum deflection recommended by the manufacturer.

3. Where PE pipes are used, a fully welded system shall be used. Mechanical or electrofusion joints shall not be used.

4. Joints shall be made in accordance with the manufacturer's instructions.

5. The number of joints shall be minimised.

6. Where, as part of the work, the protective coating or lining to the existing pipe or the new pipe is damaged, it shall be made good, as appropriate for the protective coating or lining.

5.8 WELDED JOINTS IN POLYETHYLENE PIPES

(i) Any requirement for weld tests should be described in the Contract.

1. Electrofusion and butt-fusion jointing shall be made in accordance with WIS 4-32-08, using equipment specified in WIS 4-32-16.

2. Where PE barrier or co-extruded pipes are used, the jointing system adopted shall be in accordance with the pipe manufacturer's guidance and all protective systems shall be made continuous across the joint.

3. A pipe section containing a completed weld shall achieve the same strength characteristics as the parent pipe.

5.9 FLANGED JOINTS

(i) Any special requirements for the type of flange gasket should be described in the Contract (see also Clause 2.49).

1. Jointing compounds shall not be used when making flanged joints.

5.10 OGEE JOINTS

(i) Any required jointing material (mastic or cement mortar) should be described in the Contract

1. Ogee joints shall be so made that the required jointing material fills the joint cavity. Any surplus jointing material extruded inside the barrel shall be trimmed off and, where practicable, pointed on completion.

5.11 WELDED JOINTS IN STEEL PIPES

(i) For guidance on welding, reference may be made to BS EN 1011-1, BS EN 1011-2, BS 2971, BS 4515-1 and BS 4515-2.

1. The process of welding carbon steel and stainless steel pipelines shall be in accordance with BS 4515-1 and BS 4515-2, respectively.

(ii) The types of welded joint should be described in the Contract.

(iii) The frequency and type of testing should be described in the Contract.

2. The ends of pipes shall be cut and prepared, and be free from fins, planar defects, tears and other surface defects, prior to welding. Cleaning to base metal shall extend for at least 25 mm from the end of the pipe on both internal and external faces.

3. The alignment of abutting pipe ends shall be such as to minimise the internal offset between surfaces.

4. Details of the proposed welding and welding repair procedures shall be stated in the Contract. Prior to commencement of production, test welds using these procedures shall be made under simulated Site conditions.

5. Welders shall be certified by TWI and welders shall only make welds for which they hold a current certificate.

6. Joints shall be tested using non-destructive techniques, unless it is necessary to use destructive testing to achieve adequate interpretation.

5.12 CEMENT MORTAR JOINTS

1. In making yarn and mortar joints for pipes or fittings, the spigot shall be entered into the socket of the last pipe laid until it bears on the back face of the socket and it shall be centred in the socket. Two turns of tarred yarn shall then be caulked into the back of the socket and Class M12 cement mortar shall be pressed into the joint to fill the socket and shall be bevelled off at 45° from the outside edge of the socket.

5.13 RUN LEAD JOINTS

1. Run lead joints shall be made by forcing home strands of white sterilised jute piping yarn, to the back of the socket cavity leaving a space of 75 mm (60 mm for pipes of 300 mm nominal bore and below) measured from the socket face. The socket face shall then be encircled by a suitable clip or gasket with a clay seal and the joint cavity filled with molten lead poured in one running. After cooling, the lead shall be set up and neatly finished with the face of the lead 2 mm back from the socket face. In the case of pipes over 750 mm diameter, the socket and spigot shall be heated before the joint is run.

5.14 PROTECTION OF FERROUS PIPES, JOINTS AND FITTINGS

(i) Any limitation on the type of external or internal protection required should be described in the Contract.

(ii) The type of external protection required should be described in the Contract.

(iii) For guidance on cathodic protection, see BS 7361.

(iv) Any requirements for the design of the cathodic protection should be described in the Contract.

1. Ferrous pipes, fittings and couplings shall be specified with factory-applied corrosion protection systems appropriate for their conditions of installation. Reference is made to Clause 2.36.

2. Where additional external protection is required to joints and fittings, they shall be cleaned and all loose rust removed before protection is applied. External protection shall comprise:

P1 - The application of a thin continuous coating of petroleum paste over the whole area to be protected as a primer. Where bolt heads, nuts, flanges and other projections arise, a profiling mastic shall be used to give a smooth external profile. The joint or fitting shall be wrapped with a protective tape. The minimum application shall be a spiral wrap using 55% overlap. The tape shall extend along 150 mm of the barrel of the pipe on each side of the joint or fitting.

or

P2 - The application of a continuous coating of bitumen primer over the whole area to be protected. Where bolt heads, nuts, flanges and other projections arise, a profiling mastic shall be used to give a smooth external profile. The joint or fitting shall be wrapped with a self-adhesive, cold-applied, rubber bitumen tape with a PVC backing. The minimum application shall be a spiral wrap using 55% overlap. The tape shall extend along 150 mm of the barrel of the pipe on each side of the joint or fitting.

or

P3 - The application of heat shrink sleeves.

3. Completion of internal and external protection of steel pipes shall be provided where pipes have a bituminous, epoxy or any other type of protective coating in which a gap has been left for the joint to be made. The joint and any damage to the protective coating shall be made good, see Clause 5.7.6.

4. Cathodic protection of pipes, joints and fittings shall comprise either impressed current or sacrificial anode.

5.15 CUTTING PIPES

(i) Requirements for welded joints in steel pipes are given in Clause 5.11.

(ii) 'The Control of Asbestos Regulations', 2006 the 'Control of Asbestos Regulations (Northern Ireland)' 2007, the 'Approved Code of Practices' L127 and L143 have been introduced to enable a more risk-based approach to be taken for the control of asbestos.

(iii) Guidance Notes are available from the HSE which detail the precautions to be taken when dealing with asbestos cement.

(iv) A regulatory position statement 'Leaving Decommissioned Asbestos Pipes in Excavation' and guidance notes are available from the Environment Agency.

1. Pipes shall be cut by a method which provides a clean square profile, without splitting or fracturing the pipe wall, and which causes minimum damage to any protective coating. Where necessary, the cut ends of pipes shall be formed to the tapers and chamfers suitable for the type of joint to be used, and any protective coatings shall be made good and the ends sealed in accordance with Clause 5.7.6.

2. Where ductile pipes are to be cut to form non-standard lengths, the Contractor shall comply with the manufacturer's recommendations in respect of ovality correction and tolerances to the cut spigot end.

3. Where concrete pipes are cut, any exposed reinforcement shall be sealed with an epoxy resin mortar.

4. Pre-stressed concrete pipes shall not be cut on Site.

5. Appropriate safety precautions shall be taken when cutting asbestos cement pipes.

5.16 PRECAST CONCRETE MANHOLES

(i) The type of jointing material should be described in the Contract.

1. Where steps, ladders and slabs are used in precast concrete chamber and shaft sections, they shall be aligned correctly.

2. Joints shall be made so that the required jointing material fills the joint cavity. Any surplus jointing material extruded inside the chamber or shaft shall be trimmed off and joints shall be pointed on completion.

3. Where manholes are to have a concrete surround, the concrete shall not be less than Grade GEN3 and the height of each concrete pour shall not exceed 2 m. Each construction joint shall break joint with that of the chamber or shaft sections by at least 150 mm.

4. Holes required in precast concrete chamber units to allow pipework to pass through shall be made in-situ, and oversized by the minimum amount required to ensure a watertight seal around the pipe by pre-drilling around their perimeters. Any overbreak of holes shall be made good using a mortar suitable for the pipe material.

5.17 BRICKWORK MANHOLES AND CHAMBERS

1. Manholes and chambers shall be constructed in English Bond using M12 sulphate-resisting mortar and Class B engineering bricks.

5.18 INVERTS AND BENCHING

(i) The benching material and surface finish should be described in the Contract.

(ii) The surface finishes referred to are specified in Clauses 4.21 and 4.22.

1. Where a wearing screed is required, it shall have a smooth, high-strength concrete topping applied with a steel trowel before the concrete has set.

2. Inverts and benchings in manholes, chambers and the wet well shall have a screeded, ridged finish.

5.19 PIPES AND JOINTS ADJACENT TO STRUCTURES

(i) Where a pipeline is to be constructed in ground which is variable or unstable, it may be appropriate for multiple rocker pipes to be provided to accommodate the anticipated differential settlement.

(ii) If flexible pipes are being used, rocker pipes are not needed.

1. Except where the construction is by tunnelling, heading or pipe jacking, a flexible joint shall be provided as close as is feasible to the outside face of any structure into which the pipe is built, compatible with the subsequent movement of the joint.

2. The length of the next pipe (rocker pipe) away from the structure shall be as set out below:

Nominal Diameter (mm)	Effective Length (m)
150 to 600	0.6
Over 600 to 750	1.0
Over 750	1.25

3. A pipeline may, where practicable, be laid through a manhole and the crown cut out to the half diameter, provided flexible joints are situated on each side, no further than 600 mm from the inner face of the manhole wall, and that adjacent pipes shall comply with Clause 5.19.2.

5.20 WATERTIGHTNESS OF MANHOLES AND CHAMBERS

1. Manholes and chambers shall be substantially watertight, with no discernible flow of water penetrating the Works.

5.21 SETTING MANHOLE COVERS AND FRAMES

(i) If bedding of frames on resin mortar or haunching in concrete (instead of mortar) is required, this should be described in the Contract.

1. Manhole frames shall be set to level, bedded and haunched externally over the base and sides of the frame in mortar, in accordance with the manufacturer's instructions. The frame shall be seated on at least two courses of Class B engineering bricks, on precast concrete masonry units or on precast concrete cover frame seating rings to regulate the distance between the top of the cover and the top rung to no greater than 675 mm. A mortar fillet shall be provided where the corners to an opening in a slab are chamfered and the brickwork is not flush with the edges of the opening.

2. Frames for manhole covers shall be bedded in a polyester resin mortar in all situations where covers are sited in NRSWA Road Categories I, II or III (i.e., all except residential cul-de-sacs). In all other situations, M12 mortar shall be used.

3. Installation of surface box frames and covers shall comply with the specification set out in BS 7903.

5.22 CONNECTIONS TO EXISTING SEWERS

(i) For information on repair couplings, refer to WIS 4-41-01.

1. Connections shall be made using standard pre-formed junctions where possible. Saddle connections to existing sewers shall only be allowed when the internal diameter of the major pipe is at least 150 mm greater than the internal diameter of the branch pipe.

2. Pipe saddles for concrete or clay sewers shall be bedded in Class M12 mortar and a mortar fillet formed to give a cover of at least 50 mm to the base of the saddle.

3. The ends of connections and pipes not required for immediate use shall be closed with purpose-made stoppers, disks or joiners. The position of all junctions shall be recorded by measurement from the manhole immediately downstream and recorded with the as-laid information and submitted to the Client for retention in the Client's asset records.

5.23 SEWERS AND MANHOLES TO BE ABANDONED

(i) The material required for filling should be described in the Contract.

(ii) Any requirement for clearing sewers prior to filling should be described in the Contract.

1. Where sewers are to be abandoned and filled by grouting, the lowest point of the abandoned length shall be suitably sealed, and the filling operation shall commence from that point and continue progressively so as to fill all voids completely.

2. The shafts of manholes on abandoned sewers shall be broken down to a level 1 m below finished ground level and the remaining void filled as described in the Contract.

5.24 JUNCTIONS AND LATERALS ON SEWERS

1. Junctions and laterals shall be effectively sealed with an end cap, the location of which shall be positively indicated.

2. End caps on junctions and laterals shall be secured by stakes and concrete. Marker tape shall be laid between the location peg and end cap of the lateral.

3. Records of depth and location of junctions at the sealed ends of laterals shall be recorded and a copy of the record provided to the Client as work proceeds.

5.25 MARKER AND INDICATOR POSTS

(i) Marker and indicator posts should be sufficient to allow the location of the water mains to be easily identified.

1. Marker and indicator posts shall be erected to show the location of valves and other fittings, changes of direction and at field boundaries.

2. Permanent ground markers shall be constructed at locations shown. A schedule of co-ordinates of the ground markers shall be supplied with the as-laid records and submitted to the Client for retention in the Client's asset records.

3. Marker posts shall be placed at field boundaries and in open country, and shall be at intervals of not more than one kilometre.

4. Installation of indicator plates shall comply with the specification set out in BS 7903.

5.26 TOLERANCES FOR PIPELINES

(i) Any requirements for more stringent tolerances should be described in the Contract.

(ii) Where a pipeline is to be constructed in ground which is variable or unstable, it may be appropriate for a larger tolerance for line and level to be described in the Contract.

1. The line and level of any pipeline shall not deviate from that described in the Contract by more than 20 mm and any combination of such deviations shall not create a reverse gradient.

5.27 CABLE DUCTS

(i) Clauses 2.34 and 2.91 cover materials for drawcords and cable ducting, respectively.

1. The requirements of WIMES 3.02 shall be followed.

2. Cable ducts shall be laid with a 75 mm bed and surround of sand.

3. All ducts terminating in buildings shall extend 150 mm above slab level.

5.28 INSTALLATION OF VALVES

1. During installation, valves shall be kept in the closed position.

2. Extension spindles shall be accurately installed to hold the spindle in alignment without binding. All valves shall be set so that the normal position of the operating spindles is perpendicular to the line of the main.

3. After installation, any damage to corrosion protection coatings shall be made good, in-line with Clause 5.14. Valves shall be tested to their full extent to ensure their correct operation.

4. To avoid the necessity for man entry into any chamber for valve opening/closing, an extension spindle shall be used.

5. Air valves shall be installed in drained chambers, where practical, and where this is not, they shall be installed either with their air vent(s) at a level higher than that to which water could rise or in sealed chambers.

6. Air valves shall be installed with an isolating valve between the branch/tapping off the main and themselves, to facilitate their maintenance without interruption to the mains supply.

7. Installation of fire hydrants shall comply with the specification set out in BS 9990.

5.29 WASHOUTS

(i) For provisions relating to pollution of watercourses, see Clause 1.15.

1. The washout layout shall ensure the safe disposal of the contents discharged and shall not pollute any ditch, watercourse, highway drain or surface water sewer.

2. To avoid the necessity for man entry into any chamber for operation, an extension spindle shall be used.

3. Installation of underground washouts shall comply with the specification set out in BS 7903.

5.30 ALTERNATIVE PIPE INSTALLATION TECHNIQUES

(i) 'Horizontal Directional Drilling Good Practices Guidelines (3rd Edition)' is available from the North American Society for Trenchless Technology (NASTT) (There is no guidance available in the UK).

(ii) Other drilling/moling techniques are available.

1. Pipe splitting (mains bursting) shall be carried out in accordance with 'Damage Control Procedures for Pipeline Construction Involving Pipe Splitting' issued jointly by British Gas (currently Transco) and UKWIR.

2. Directional drilling shall be undertaken in one continuous operation.

3. The drill/mole shall be fitted with a signal generator to assist in locating the end and the route of the drill

5.31 CONNECTIONS TO EXISTING WATER SUPPLY PRESSURE PIPELINES

1. The dimensions of the existing pipe at the point of connection shall be checked to ensure the compatibility of the proposed connection fittings.

5.32 TEMPORARY WATER SUPPLY MAINS

(i) The arrangements for the maintenance and removal of temporary water supply mains should be detailed in the Contract.

(ii) Where temporary water supplies are required to cross highways and roads, prior approval may be required from the relevant Highway Authority. Arrangements for the protection of these temporary mains should be detailed in the Contract.

1. The sizes, materials and locations of all temporary mains shall be agreed with the Client, prior to commencement. All temporary mains shall be suitable for the working pressure in the area.

2. Temporary mains shall be set up, disinfected and sampled, as specified by the Client. The relevant connections shall be made and supply restored within a period agreed with the Client.

3. The Contractor shall ensure that all temporary hoses and temporary mains are kept free from ice and are fully operational during cold weather.

5.33 WATER MAINS TO BE ABANDONED

(i) A regulatory position statement 'Leaving Decommissioned Asbestos Pipes in Excavations' and guidance notes are available from the Environment Agency.

(ii) The fill material should be detailed in the Contract.

1. Where water mains or services are abandoned, they shall be physically severed from the network. A closed gate valve shall not be used to achieve disconnection.

2. Where water mains are to be abandoned and filled by grouting, the lowest point of the abandoned length shall be suitably sealed, and the filling operation shall commence from that point and continue progressively so as to fill all voids completely.

3. All abandoned mains shall be identified and recorded with the as-laid information and submitted to the Client for retention in the Client's asset records.

4. Asbestos cement mains, where decommissioned, shall be recorded with the as-laid information and submitted to the Client for retention in the Client's asset records. Where an asbestos main has been burst and asbestos shards remain in the ground, this shall be recorded with the as-laid information and submitted to the Client for retention in the Client's asset records.

SECTION 6

BUILDING WORKS

The Notes for Guidance are not part of the Specification.

(i) This Section is intended only for small-scale masonry buildings and basic steel framed and clad enclosures on operational sites. Where design is included in the Contract, all buildings which are occasionally occupied and/or heated should, as a minimum, comply with the Building Regulations 2000.

6.1 BRICKWORK AND BLOCKWORK GENERALLY

(i) Any requirement for rendering of manholes and chambers should be described in the Contract.

(ii) The bond should be described in the Contract.

(iii) The required class of mortar and type of cement should be described in the Contract (see Clause 2.82).

(iv) For guidance on the design of joints and jointing in building construction, see BS 6093.

1. Brickwork and blockwork shall comply with the relevant provisions of BS EN 1996-2.

2. The moisture content of the bricks shall be adjusted so that excessive suction is not exerted on the mortar.

3. Bricks in each course shall break joint correctly with the bricks underneath. The courses shall be laid parallel with joints of uniform thickness and shall be kept straight or regularly curved, as required. Brickwork shall be gauged to rise 300 mm in four courses. Vertical joints shall be in alignment, as required by the bond, and shall have an average thickness of 10 mm. Bricks forming reveals and internal and external angles shall be selected for squareness and built plumb. Bricks with single frogs shall be laid frog upwards.

4. Brickwork and blockwork shall rise uniformly; corners and other advanced work shall be raked back and not raised above the general level more than 1 m. No brickwork shall be carried up higher than 1.5 m in one day. No bats or broken bricks shall be incorporated in the work, unless essential for bond. Where cut blocks are required, all cutting shall be carried out with a mechanical cutting disk.

5. Completed brickwork and blockwork shall be protected at all times from scaffold splash, mortar droppings, grout leakage from suspended slabs and the harmful effects of weather. Brickwork and blockwork shall be allowed to set thoroughly hard before cutting or chasing is carried out.

6. Blocks used in fair faced blockwork shall have a smooth undamaged face and be neatly pointed with flush joints.

6.2 BRICKWORK AND BLOCKWORK, JOINTING AND POINTING

(i) The type of jointing and pointing should be described in the Contract.

(ii) Any requirement for sample panels and their approval should be described in the Contract.

1. Bricks and blocks shall be laid in mortar, properly bedded and jointed, and all joints filled with mortar at every course.

2. Where the surface of walling does not provide an adequate key, the joints on faces of walls to be plastered shall be raked out 12 mm deep.

6.3 CAVITY WALLS

(i) Where the filling or semi-filling of cavities with insulating material is required, it should be described in the Contract.

1. Cavity walls shall have a minimium cavity width of 50 mm and a minimal air gap of 25 mm between any cavity insulation and the outer skin. Cavity walls shall be built with wall ties uniformly spaced, 450 mm apart vertically and 750 mm apart horizontally, staggered, and laid to fall outwards. Additional ties shall be used near the sides of all openings, one for each third course of bricks. Care shall be taken to keep the ties within the cavity free from mortar or mortar droppings, and any mortar or debris collecting at

the bottom of the cavity shall be cleaned out through temporary openings left for this purpose in the bottom courses.

2. Weepholes shall be provided by leaving open perpendicular joints at intervals not greater than 900 mm in the course immediately above the cavity tray. Where the cavity tray does not extend the full length of the exposed wall, i.e., at an opening and stop end, at least two weep holes, not more than 450 mm apart, shall be provided.

6.4 DAMP-PROOF COURSES

1. Construction of damp-proof courses shall comply with the relevant provisions of BS EN 1996-2, BS 6576, BS 8102 and BS 8215.

6.5 CORBELLING

1. Oversail corbelling shall not exceed 30 mm on each course.

6.6 BONDING TO CONCRETE

1. Where brickwork or blockwork is to be bonded to concrete, this shall be achieved by means of metal ties evenly placed at three per square metre. Brickwork and blockwork shall be brought up subsequent to the concrete.

6.7 UNDERPINNING

(i) Larger-scale underpinning should be described in the Contract and may require the services of a specialist contractor.

1. Underpinning in brickwork shall be carried up to within one course of the underside of the existing structure and allowed to set. The remaining course shall be bedded in mortar and wedged tightly against the existing structure with slate pieces.

6.8 CENTERING AND LAGGING

1. Centering and lagging used for the construction of brickwork and blockwork shall remain in place for such time as is necessary for the brickwork and blockwork to develop sufficient strength to prevent sagging or cracking of joints.

6.9 BRICKLAYING AND BLOCKLAYING IN COLD WEATHER

1. Materials used in bricklaying and blocklaying shall be frost free and no bricks or blocks shall be laid when the ambient temperature is below 3°C, unless special precautions are taken. Any additives used in the mortar shall not cause a variation in the colour of the joints. Completed work shall be protected adequately during cold weather.

2. General working practices shall be in accordance with the Brick Development Association publication 'Bricklaying in Winter Conditions'.

6.10 PREPARATION FOR PLASTERING

(i) The type of scrim material should be described in the Contract.

(ii) If joints are to be cut or covered as an alternative to scrim, this should be described in the Contract.

1. Unless a bonding agent is used, concrete ceilings, ceiling beams, columns and stanchions shall be dubbed out, as necessary, before plastering is commenced and the mix used for dubbing shall be similar to that used for first undercoating. The surface of in-situ concrete shall be cleaned of dust, loose particles and other matter. Surfaces shall be wetted immediately before plastering is commenced.

2. Angles between walls and ceilings, vertical angles and joints between dissimilar solid backgrounds, shall be reinforced with 90 mm wide scrim set in plaster and trowelled flat. All joints between plasterboards shall be similarly treated.

3. Expanded metal angle beads shall be provided at all external corners.

6.11 FIXING OF PLASTERBOARD

1. Drylining shall comply with BS 8212. Plasterboard for ceilings shall be fixed to the support at 150 mm centres with plasterboard nails, and fastened so that the joints are staggered. Noggins or other fixing surface shall be provided, as necessary, to ensure that edges of plasterboard are secured adequately. Ends of sheets shall be butted tightly and edges left with a gap not exceeding 5 mm.

2. Where sheeting has been cut, nails shall not be less than 18 mm from cut edges. Nails shall be driven well home, with heads slightly below the surface, but shall not break the paper.

6.12 PLASTERING

1. Plastering shall normally be applied in two coats, in accordance with the manufacturer's instructions, and batches shall be used as soon as possible after water has been added. The total thickness of both coats shall be of the order of, but shall not exceed, 13 mm.

2. Where three coat work is described in the Contract, the thickness shall be of the order of, but shall not normally exceed, 18 mm.

3. The thickness of two coat work applied to concrete ceilings and soffits or plasterboard shall not exceed 9 mm.

4. The thickness of finishing coats shall be of the order of 3 mm, except where board finished plaster is used when the thickness shall be 5 mm and comply with the relevant provisions of PD CEN/TR 15123, BS EN 13914-2 and BS 8481.

5. Expanded metal beads, or similar, shall be provided at all arrises.

6.13 PLASTERING IN COLD WEATHER

1. When the ambient temperature is 5°C or less, the portion of the Works to be plastered shall be completely enclosed. The ambient temperature shall be raised and maintained above 5°C until the completion of plastering and hydration.

6.14 CONCRETE FLOOR FINISHES

(i) The type of finish from BS 8204-2 and BS 8204-4 should be described in the Contract.

1. Concrete floor finishes shall comply with the relevant provisions of BS 8204-2 and BS 8204-4.

2. Minimum thickness of sand/cement floor screeds shall be 35 mm when laid directly on a prepared concrete slab or 75 mm when separated from the slab by a membrane.

3. Structural concrete slabs shall be suitably treated to provide a full bond prior to screeding, unless a membrane is to be used.

4. Screeds laid in wash down areas shall be laid to a minimum fall of 1 in 40.

6.15 FLOOR TILING

1. Floor tiling shall comply with the relevant provisions of BS 5385-3 and BS 5385-5 for rigid tiling and BS 8203 for flexible tiling.

6.16 TERRAZZO FLOOR FINISHES

1. Terrazzo floor finishes shall comply with the relevant provisions of BS 8204-4.

6.17 EXTERNAL RENDERING

1. External rendering shall comply with BS EN 13914-1 and shall be applied to a total thickness of not less than 20 mm. The mix for both coats shall be as for Class M2 mortar. The first coat shall be applied, levelled, scratched and left to dry for not less than 3 days in warm weather and not less than 7 days in cold or wet weather. The suction of the surface of the first coat shall be adjusted, as necessary, by wetting before applying the second coat which shall be coloured as directed, levelled and lightly trowelled with a wooden trowel.

6.18 WALL TILING

1. Wall tiling shall comply with the relevant provisions of BS 5385-1, BS 5385-2 and BS 5385-4.

6.19 CARPENTRY AND JOINERY

1. Wherever possible, cutting and shaping of all timber shall be completed before preservative treatment is carried out. Where any cutting or shaping has to be carried out after treatment, the cut or worked surfaces shall be given two coats of the preservative. After treatment, timber shall be thoroughly dried out before use.

2. The whole of the joinery shall be cut and framed together as soon as possible after the commencement of the work. Workmanship shall comply with the relevant provisions of BS 1186-2 and BS 6446. Except where work is described in the Contract as being to finished sizes, 3 mm shall be allowed for each wrot face. Frames, casings and other joinery fittings shall be secured to hardwood fixing slips, built in for the purpose. Where hardwood fixing slips have not been provided, receiving surfaces shall be plugged with hardwood plugs or other suitable type plugs.

3. Manufactured units to be painted shall be primed at the place of manufacture.

6.20 STRUCTURAL STEELWORK

(i) This clause is intended to apply to basic structural steel framed buildings, used to enclose process equipment.

(ii) Any particular design requirements or architectural finishes should be described in the Contract.

(iii) Where steelwork is to be galvanised, it should be described in the Contract, unless covered by Clause 6.20.3.

(iv) Other protective coatings, if required, should be described in the Contract.

(v) Requirements for the testing of welds should be described in the Contract.

(vi) Structural steelwork should comply with the 'National Structural Steelwork Specification for Building Construction' published by the British Constructional Steelwork Association.

(vii) The Contract should state the arrangements for sampling and testing of all surface treatment materials.

1. Built in joists used in masonry structures shall be cut to length by cold sawing. The minimum bearing length on masonry shall be 100 mm, and minimum crushing strength of supporting blockwork shall be 7 N/mm^2 or a padstone provided to spread the load.

2. Structural steelwork for steel framed buildings shall be designed, detailed, fabricated and erected in accordance with BS EN 1993-1 and BS EN 1090-2.

3. All structural steelwork in buildings housing wet processes, or otherwise prone to condensation, shall be galvanised to BS EN ISO 1461 with a coat thickness of 85 µm.

4. Welders employed in the fabrication of steelwork shall be certified to BS EN 287-1 for tests appropriate to the work to be undertaken.

5. Steelwork to be embedded in concrete shall have any loose mill scale and rust removed.

6. As a minimum, all steelwork to be coated shall be blast cleaned to second quality, then treated with two coats of quick drying primer, reference shall be made to BS 7079. All surface treatment shall comply with the relevant provisions of BS EN ISO 12944 and shall be completed, as far as practicable, before steelwork is delivered to Site. Sampling and testing of all surface treatment materials shall be in accordance with the Contract.

7. All steelwork fastenings shall receive protective treatment to at least as good a standard as the parts connected.

8. After tightening, all bolts shall have at least two threads projecting beyond the nut or lock nut. Bolt shank length shall be such as to ensure at least one thread shall remain clear between the nut and the shank.

9. When required, the following tests of butt welds shall be made;

(a) Ultrasonic examination to BS EN 1714;

(b) Penetration test to BS EN 571-1; or

(c) Magnetic particle test to BS EN ISO 9934-1.

10. Erection tolerances for buildings containing an overhead gantry crane should comply with the requirements of Appendix F in BS 466:1984.

6.21 ROOFS

1. Roof members shall be fabricated in accordance with BS EN 1995-1-1.

2. Flats and gutters shall be covered with BS EN 636-3G plywood or tongued and grooved wrot boarding laid diagonally, firred to falls of not less than 1 in 120 for lead and copper and 1 in 60 for bitumen felt.

6.22 TIMBER FLOORS

(i) Thickness of floor boarding should be described in the Contract.

(ii) If chipboard is to be used instead of floor boarding, this should be described in the Contract.

1. Floor joists shall be either built into brickwork or blockwork, or held in galvanised steel joist hangers, and shall be trimmed, as described in the Contract. Bridging shall be spaced at every 1.8 m apart and shall be 50 mm thick to the full depth of the joists, or 38 mm by 50 mm herringbone strutting.

2. Boarding shall be cramped up and nailed with cut flooring nails. Trimmed openings shall have mitred borders 75 mm in width.

6.23 DOOR FRAMES

1. Door frames shall be fitted into prepared openings, drilled and plugged at three points per jamb. Fixings shall be suitable for the proposed application and be not more than 600 mm apart and not more than 150 mm from top or bottom. The door frame shall be bedded in non-shrink or epoxy grout.

2. The joint between external door frames and adjacent walls shall be continuously pointed with gun-applied butyl or other approved non-setting mastic.

6.24 WINDOWS

(i) Security requirements for windows and doors should be stated in the Contract.

1. Window frames shall be securely fixed to openings, in accordance with manufacturer's instructions, and shall be continuously pointed with gun-applied butyl or other suitable non-setting mastic.

6.25 GLAZING

1. Glazing shall comply with the relevant provisions of BS 6262.

6.26 PAINTING

(i) BS EN 927-1 gives guidance on exterior paint protection systems.

(ii) Refer to Clause 2.88 for paintwork on buildings.

1. Painting of structural steelwork shall comply with the relevant provisions of BS EN ISO 12944. Other painting shall comply with the relevant provisions of BS 6150.

6.27 SLATING AND TILING

(i) The types and sizes of slate and tiles, together with details of any laps, battens, fixings, beddings, underfelt and boarding, should be described in the Contract.

1. Slating and tiling work shall comply with the relevant provisions of BS 5534.

6.28 LIGHTWEIGHT CONCRETE ROOF SCREEDS

1. Lightweight concrete roof screeds shall be laid to the necessary falls and shall comply with the relevant provisions of BS 8217. Before any felt is laid, a coat of bitumen primer shall be applied and allowed to dry.

6.29 ASPHALT ROOFING

(i) A written guarantee may be available for asphalt work incorporating terms currently applied by the Mastic Asphalt Council and Employers' Federation.

1. Mastic asphalt shall be laid in accordance with the relevant provisions of BS 8218 on an underlay of sheathing felt, laid loose and with joints lapped at least 50 mm. The roof shall be set out with properly formed high points, water lines and mitred bays, where required.

6.30 BITUMEN FELT ROOFING

(i) The surface finish should be described in the Contract.

1. Built-up bitumen felt roof coverings shall be laid in accordance with the relevant provisions of BS 8217.

6.31 PLUMBING

(i) Where heating and hot water systems are to be installed, reference should be made to BS EN 12828, BS EN 12831 and BS EN 14336.

(ii) Reference should be made to the Water Supply (Water Fittings) Regulations 1999, the Water Supply (Water Fittings) Regulations (Northern Ireland) 2009 or the Scottish Water Byelaws 2004.

1. Plumbing for domestic water supply shall comply with the relevant provisions of BS 6700, BS EN 806-2 and BS EN 806-4.

6.32 OPENINGS IN WALLS, FLOORS AND CEILINGS

1. Box outs and/or cut openings through walls, floors and ceilings for the passage of pipes and cables shall be provided. Where described in the Contract, tube sleeves cut off flush with the finished surface shall be provided and fixed in position. All openings and ducts shall be sealed on completion to prevent the passage of liquids, or toxic or explosive gases.

6.33 TOLERANCES FOR BUILDING WORKS

1. Tolerances for building works, except where otherwise described in the Contract, shall not exceed the permissible deviations from levels and dimensions given in BS 5606, where applicable, for the corresponding types of work.

6.34 ELECTRICAL INSTALLATIONS

1. All electrical installations shall comply with the 'Requirements for Electrical Installations', BS 7671 published by the Institution of Electrical Engineers.

6.35 PROFILED STEEL CLADDING

(i) The colour of profiled steel may require the approval of the Local Planning Authority.

1. Profiled steel roof cladding shall be designed in accordance with BS EN 1993-1-3. In addition, the combination of purlin spacing and sheeting profile shall be of adequate stiffness to prevent undue deflection (such as to cause loss of balance) when walked on during erection and maintenance.

2. Sheets shall be supplied to Site, cut to the required length. Any incidental Site cuts shall be sealed in accordance with the manufacturer's instructions, using matching materials.

3. Any cladding sheets with damage to the external protective coating shall be rejected. Other damage shall be repaired as in Clause 6.35.2. Sheeting shall be carefully stacked on Site with its dust wrap in place, until required.

4. Flashings and fillers shall be purpose-made to suit the profile of the sheets being used.

6.36 PRECAST CONCRETE FLOORS

(i) These clauses are a guide for the use of "beam and block" or pre-stressed concrete planks up to 5 m span.

1. Units shall be delivered cut to the correct lengths and complete with any service openings required. Any Site modifications shall be carefully cut in accordance with manufacturer's instructions, following specific permission from the manufacturer in each case.

2. Any unit with handling damage to the bearing area shall be rejected.

3. Minimum bearing areas for precast units bearing on steelwork shall be as laid down by the manufacturer. No negative tolerance shall be allowed on these dimensions. Units bearing on masonry shall have a minimum bearing length of 100 mm and the strength of the masonry shall be as specified by the designer, but not less than 7 N/mm^2.

4. In-situ structural topping shall be cast in the sequence laid down by the Designer. Propping requirements shall be exactly followed and props left in place until the specified strength is reached.

5. Bricks, blocks and other materials shall not be stored on the floor.

6. Units with excessive pre-camber resulting in differential levels between planks exceeding 10 mm shall be rejected. Screeded floor finishes shall be to the tolerances laid down in Clause 4.25.

6.37 COMPOSITE FLOORS

1. In-situ concrete floors utilising permanent steel soffit formwork as part of a composite structural construction, shall not be used in buildings housing wet processes or otherwise subject to condensation.

2. The exposed soffit of any such permanent formwork used shall be protected to the same standard as the general structural steelwork.

SECTION 7

TESTING AND DISINFECTION

The Notes for Guidance are not part of the Specification.

(i) The Contract should describe requirements for:

(a) temporary pipework, its connection to existing mains and associated testing apparatus;

(b) disinfection, chemical and bacteriological sampling, and the test parameter values to be achieved; and

(c) samples and/or frequency of testing of materials.

(ii) Refer to Guidance Note 2 of the Water UK 'Principles of Water Supply Hygiene' for guidance on the hygienic transport and storage of new potable water pipes.

(iii) Where water fittings and appliances are to be used to convey water derived from the public water supply, which brings them under the scope of the Water Supply (Water Fittings) Regulations 1999, the Water Supply (Water Fittings) Regulations (Northern Ireland) 2009 or Scottish Water Byelaws 2004, pressure testing should be undertaken in accordance with these regulations.

7.1 CLEANSING AND SWABBING OF PIPELINES

(i) The pipeline cleaning system should be appropriate for the pipe material and should be described in the Contract.

(ii) The Contract should describe the type of swabs and the number of passes.

1. On completion of construction, and before any disinfection, internal surfaces of pipelines shall be cleaned thoroughly.

2. On completion of the hydraulic test on water mains, a foam swab shall be passed through the main a sufficient number of times to achieve clear wash water.

3. All swabs shall be recovered and accounted for after mains cleaning.

7.2 PRECAUTIONS PRIOR TO TESTING PIPELINES

(i) See also Clauses 5.1 and 5.6.

1. Before testing any pipeline, it shall be anchored adequately and thrusts from bends, branch outlets or from the pipeline ends shall be transmitted to solid ground or to a suitable temporary anchorage.

2. Open ends shall be stopped with plugs, caps or blank flanges, properly jointed.

3. Testing against closed valves shall not be permitted.

7.3 TESTING METHOD PROGRAMME AND NOTIFICATION

1. The proposed programme and method of testing and swabbing shall be stated in the Contract.

2. The Client shall be notified of the intention to test a section of pipeline within the period specified in the Contract.

7.4 TESTING NON-PRESSURE PIPELINES

(i) The type of test (air, water, visual or CCTV) should be described in the Contract.

1. Non-pressure pipelines laid in open cut shall be tested after they are jointed and before any concreting or backfilling is commenced, other than such as may be necessary for structural stability whilst under test.

2. The pipelines shall be tested by means of an air or water test or by a visual or closed-circuit television (CCTV) examination, in lengths determined by the course of construction, in accordance with the programme.

3. A further test shall be carried out after the backfilling is complete.

7.5 WATER TEST FOR NON-PRESSURE PIPELINES

(i) For provisions relating to pollution of watercourses, see Clause 1.15.

(ii) The allowable quantity of water added is stipulated in Clause 13.3 of BS EN 1610:1998.

1. The test pressure for non-pressure pipelines, up to and including 1000 mm nominal bore, shall be not less than 1.0 m head of water above the pipe soffit or groundwater level, whichever is the higher at the highest point, and not greater than 5 m head at the lowest point of the section. Steeply-graded pipelines shall be tested in stages in cases where the maximum head, as stated above, would be exceeded if the whole section were tested in one length.

2. The pipeline shall be filled with water and a minimum period of 1 hour shall be allowed for absorption, following which the original water level shall be restored. Water shall then be added from a measuring vessel at intervals of 5 minutes over a 30-minute period and the quantity required to maintain the original water level noted. The length of pipeline shall be accepted if the quantity of water added in 30 minutes does not exceed 0.15 litres/m^2 for pipelines or 0.2 litres/m^2 for pipelines and manholes tested together, where m^2 refers to the total area of the wetted internal surface.

3. Following a water test, pipelines and structures shall be emptied, as far as is practicable.

7.6 AIR TEST FOR NON-PRESSURE PIPELINES

(i) BS EN 1610 'Construction and Testing of Drains and Sewers', gives advice on testing requirements.

1. Non-pressure pipelines to be air tested shall have air pumped in by suitable means until a pressure of 100 mm head of water is indicated in a U-tube connected to the system. The pipeline shall be accepted if the air pressure remains above 75 mm head of water after a period of time "t" given in the following Table without further pumping, following a period for stabilisation.

Nominal Diameter	Testing time (minutes)
DN 100	5
DN 200	5
DN 300	7
DN 400	10
DN 600	14
DN 800	19
DN 1000	24

2. Failure to pass the test shall not preclude acceptance of the pipeline if a successful water test can subsequently be carried out in accordance with Clause 7.5.

7.7 CCTV INSPECTION OF PIPELINES

(i) CCTV inspection should be described in the Contract, if required.

1. CCTV inspections shall comply with the 'Model Contract Document for Sewer Condition Inspection' 1994, published by The Foundation for Water Research.

7.8 INFILTRATION

(i) The permissible infiltration is the same as the permissible loss in the water test in Clause 7.5.

1. Non-pressure pipelines (including tunnels and shafts) and manholes shall be tested for infiltration after backfilling. All inlets to the system shall be effectively closed, and any residual flow shall be deemed to be infiltration.

2. The pipeline, including manholes, shall be accepted as satisfactory if the infiltration, including infiltration into manholes, in 30 minutes does not exceed 0.2 litres/m^2, where m^2 refers to the total internal surface area of the pipeline, including manholes.

3. Notwithstanding the satisfactory completion of the above test, if there is any discernible flow of water entering the pipeline at a point which can be located either by visual or CCTV inspection, then measures as are necessary to stop such infiltration shall be taken.

7.9 TESTING OF DUCTILE IRON, PVC, GRP AND STEEL PRESSURE PIPELINES

(i) BS EN 805 gives the permissible losses during the test.

(ii) The Contract should specify either a "water loss" or "pressure loss" method of testing in accordance with BS EN 805.

(iii) Details for Site pressure testing in accordance with BS EN 805 are provided in IGN 4-01-03 'Field Pressure Testing of Pressure Pipes and Fittings'.

(iv) Where a pipeline is to be tested in sections, the Contract should identify any constraints in length.

(v) The test pressures should be described in the Contract.

(vi) Testing should be carried out with the pipework isolated from the existing adjacent pipework to prevent the transmission of thrust.

1. The entire pipeline shall be pressure tested in accordance with BS EN 805.

2. Gauges used for testing pressure pipelines shall either be of the conventional circular type, not less than 200 mm diameter, calibrated in metres head of water, or shall have a digital indicator capable of reading increments of 0.1 m head. Before any gauge is used, it shall be checked independently and a dated certificate of its accuracy shall be provided.

3. Before testing, valves shall be checked and sealed, the sections of main filled with water and the air released. After having been filled, pipelines shall be left under normal operating pressure for 24 hours, so as to achieve conditions as stable as possible for testing.

4. The pressure in the pipeline shall then be raised steadily until the specified test pressure is reached in the lowest part of the section, and the pressure shall be maintained at this level, by pumping, if necessary, for a period of one hour. The pump shall then be disconnected, and no further water shall be permitted to enter the pipeline for a further period of one hour. At the end of this period, the original pressure shall be restored by pumping and the loss measured by drawing off water from the pipeline until the pressure as at the end of the test is again reached.

5. Where a new pipeline is to connect to an operational pipeline, the final connection shall be inspected visually under normal operating pressure and there shall be no visible leakage.

7.10 TESTING OF POLYETHYLENE PRESSURE PIPELINES

(i) Where a pipeline is to be tested in sections, the Contract should identify any constraints in length.

(ii) The test pressures should be described in the Contract.

(iii) Testing should normally be carried out with the pipework isolated from the existing adjacent pipework to prevent the transmission of thrust. However, this might not be appropriate where the test is designed to test the tension capability of welds and the transfer of thrust through the pipeline.

1. The testing of polyethylene pressure pipelines shall be carried out in accordance with the procedures in IGN 4-01-03 'Field Pressure Testing of Pressure Pipes and Fittings'.

2. Clauses 7.9.2 and 7.9.5 shall apply.

7.11 DISINFECTION OF WATER MAINS

(i) The Contract should describe the sampling parameters.

(ii) Any specific requirements for bringing water mains back on line should be included in the Contract.

(iii) Timescales for the return of sample results from bacteriological and chemical sampling should be described in the Contract.

(iv) Refer to Clause 1.16 for details relating to water supply hygiene.

1. After satisfactory completion of testing and, if applicable, after swabbing, pipelines intended for conveying potable water shall be disinfected in accordance with 'Principles of Water Supply Hygiene' and the accompanying Guidance Notes published by Water UK, or as specified in the Contract.

2. After disinfection and bacteriological and chemical sampling of completed sections of water mains, no valves shall be turned or any other action taken which might interfere with the main or existing network without the prior agreement with the Water Undertaker.

3. On completion of the disinfection, the pipeline shall be left full of water under operating pressure and with sufficient flow or changes of water so as to maintain water quality.

4. Groundwater shall not be allowed to come into contact with the inside of water mains after disinfection.

7.12 CLEANSING OF STRUCTURES

1. On completion of construction, and before any disinfection, internal surfaces of structures designed to retain or convey an aqueous liquid shall be cleaned thoroughly in such a way as to remove all oil, grit and other deleterious matter.

7.13 TESTING OF CONCRETE ROOFS

(i) The Contract should describe which of the two alternative tests is required. Lagooning is the preferred method but hosing may be used for smaller structures.

1. Concrete roofs of structures which are to contain aqueous liquids shall be watertight and shall, where practicable, be tested prior to the installation of any waterproof membrane by lagooning with water to a minimum depth of 25 mm for a period of 24 hours.

2. Where it is impracticable because of roof falls, or otherwise, to contain 25 mm depth of water, the roof shall be thoroughly wetted by continuous hosing for a period of not less than 6 hours.

3. In either case, the roof shall be regarded as satisfactory if no leaks or damp patches show in the soffit.

4. The roof covering shall be completed as soon as possible after satisfactory testing.

7.14 TESTING OF CONCRETE STRUCTURES DESIGNED TO RETAIN AN AQUEOUS LIQUID

1. After cleaning, and as far as practicable before any earth or other filling is placed against the outside wall faces, concrete structures designed to retain an aqueous liquid shall be filled with water at a uniform rate of not greater than 2 m in 24 hours. A period of up to 21 days shall be allowed for stabilisation, after which the water level shall be recorded at 24-hour intervals, for a test period of 7 days. During the test period, the total permissible drop, after allowing for evaporation and rainfall, shall not exceed 1/500 of the average water depth of the full tank or 10 mm, whichever is the lesser.

2. If the requirements for the 7-day test are not met then, after completion of any remedial work, the structure shall be refilled and, if necessary, left for a further stabilisation period after which the water level shall be recorded at 24-hour intervals, for a test period of 7 days.

2. Not withstanding the satisfactory completion of the above test, any leakage visible on the outside faces of the structure shall be stopped. Any caulking or making good of cracks in the wall section shall, where practicable, be carried out from the inside face.

3. The hydraulic test shall be carried out after successful completion of the roof tests, before any disinfection work and before any sealing is undertaken.

4. Adjacent internal chambers within a structure shall be tested sequentially. Chambers adjacent to the chamber under test shall be empty during the test.

5. On satisfactory completion of the test, the structure shall be emptied, as far as practicable, unless the water can be used as part of subsequent activities.

7.15 DISINFECTION OF STRUCTURES FOR POTABLE WATER

(i) This Clause will not apply if the Client directly undertakes disinfection.

(ii) For provisions relating to pollution of watercourses, see Clause 1.15.

(iii) The Contract should describe the sampling parameters.

(iv) Responsibility for taking samples should be described in the Contract.

(v) Timescales for the return of sample results from bacteriological and chemical sampling should be described in the Contract.

(vi) The period within which the Client should make the sample test results available to the Contractor should be stated in the Contract.

(vii) The advance period notice for taking of samples should be stated in the Contract.

1. Immediately before acceptance of any structure for potable water, the interior shall be disinfected using water chlorinated to give a residual of not less than 20 mg of free chlorine per litre. Roofs and other surfaces above top water level shall be sprayed with a solution containing 20 mg chlorine per litre. The structure shall be emptied, flushed with mains water and then filled with water having a chlorine residual of not more than 0.5 mg free chlorine per litre to normal top water level.

2. After 24 hours, samples shall be taken for bacteriological and chemical analysis. The structure shall be deemed to be disinfected if there are no coliforms in the sample.

3. The Client shall arrange for samples to be taken which are representative of the bulk of the water in the structure.

4. On completion of the disinfection, the structure shall be left full of potable water, under operating pressure and with sufficient flow or changes of water so as to maintain water quality.

7.16 WATER FOR TESTING, SWABBING AND DISINFECTION

(i) The Contract should describe any constraints relating to the temporary supply, including flow rate, pressure, location of the supply and period of notice required to make supplies available.

(ii) Any particular requirements for bridging pipework should be described in the Contract.

(iii) The arrangements for recording and reporting volumes of water used for testing, swabbing and disinfection should be stated in the Contract.

1. Water for testing, swabbing and disinfection of potable water mains and structures shall be taken from the existing potable water supply. Arrangements shall be made with the Client for appropriate supply facilities.

7.17 DISPOSAL OF WATER FROM CLEANSING, TESTING OR DISINFECTION

(i) Flow rates should be specified in the Contract for discharges to sewers.

(ii) Consent to discharge to a watercourse is issued by the Environment Agency in England and Wales, and the Scottish Environmental Protection Agency in Scotland.

1. The provisions for the removal and disposal of water used for disinfection, swabbing or testing shall be stated in the Contract.

2. Discharges to sewers shall not take place without the consent of the Sewerage Undertaker.

3. Water used in the cleansing, testing or disinfection of structures or pipelines shall be rendered safe prior to discharge to the environment.

(iii) Water from disinfection is high in chlorine and may have an adverse effect on the environment and the operation of sewage treatment works.

(iv) For provisions relating to pollution of watercourses, see Clause 1.15.

7.18 TESTING OF NON-CONCRETE STRUCTURES FOR RETAINING AQUEOUS LIQUIDS

(i) Tests for non-concrete roofs should be described in the Contract.

(ii) The method for assessing the change in water level should be described in the Contract, together with a reasonable interpretation of "no discernible change".

(iii) Rainfall and evaporation could be assessed by using a test tank of a similar size.

1. After cleaning, and before any external fill is placed against the structure, it shall be filled to its overflow level at a rate approved by the manufacturer, or detailed of the structure in the Contract. The water level shall then be recorded for a period of 3 days. The tank shall be deemed to have passed the test if there is no discernible change in water level after taking account of rainfall and evaporation, and that there are no visible signs of leakage from the walls or base of the structure.

2. This test shall only be carried out after a successful test of any roof structure.

3. Internal compartments shall be tested separately. Compartments adjacent to the compartment under test shall be empty during the test.

SECTION 8

ROADWORKS

The Notes for Guidance are not part of the Specification.

(i) This Section is not intended to cover the construction of prospectively adoptable highways but, together with the relevant Materials Clauses, is based generally on the requirements of the 'Specification for Highway Works', published by the Highways Agency, abbreviated and simplified to take account of the type of road normally required in association with Water Industry work.

8.1 ROAD FORMATIONS

(i) A cross-section of the carriageway construction and the position and type of any joints, should be described in the Contract.

1. The road formation shall be the surface obtained after completion of any earthworks.

2. Formations, immediately before being covered with sub-base or road base material, shall be clean, free from mud and slurry and properly shaped and compacted to an even and uniform surface.

3. The preparation and surface treatment of formations shall be carried out after the reinstatement of any excavations for services.

4. The formation shall be compacted so that the dry density of the upper 150 mm of the ground or fill is less than 95% of the maximum dry density, determined in accordance with Section 4 of BS 1377-4. Where the formation is in natural ground, the compaction shall, whenever possible, be carried out at or near the natural moisture of the ground.

8.2 SUB-BASE CONSTRUCTION

(i) The use of approved recycled materials is actively encouraged.

(ii) If a lean-mix concrete sub-base is required, this should be described in the Contract.

1. Within 48 hours of completion of a road formation, granular sub-base material shall be spread and compacted to the required thickness. The sub-base shall be protected from deterioration due to ingress of water, the adverse effects of weather and the use of construction equipment. Compaction shall be carried out in accordance with the relevant Highway Reinstatement Specification.

8.3 WET-MIX MACADAM CONSTRUCTION

1. Wet-mix macadam shall be spread evenly on the sub-base and compacted in layers of not more than 200 mm thickness at the optimum moisture content ± 0.5%.

2. Spreading shall be undertaken concurrently with placing. Compaction shall be completed as soon as possible after the material has been spread and carried out in accordance with the relevant Highway Reinstatement Specification.

8.4 LEAN CONCRETE CONSTRUCTION

1. Lean concrete construction for roads shall be GEN1 and shall be spread evenly on the sub-base and laid and compacted in layers of not more than 200 mm thickness.

2. Spreading shall be undertaken concurrently with placing. Compaction shall be completed as soon as possible after the material has been spread and carried out in accordance with the relevant Highway Reinstatement

Specification. The maximum period of time between mixing of the materials and final compacting of any given material shall be 2 hours.

3. Where practicable, joints against hardened material shall be avoided. Where such joints are unavoidable, the hardened material shall be cut back vertically for the full depth of the layer before placing any further adjacent material.

4. Lean concrete shall be cured for a period of not less than 7 days. No vehicular traffic shall be allowed to run on the base during this time.

8.5 LAYING COATED MACADAM

1. Transportation, laying and compaction of all coated macadam shall be carried out in accordance with the relevant provisions of BS 594987.

2. Where laying against an exposed edge, the joint shall be treated with an application of bitumen conforming to BS EN 12591 Grade 100/150.

8.6 LAYING HOT ROLLED ASPHALT

1. Transportation, laying and compaction of hot rolled asphalt shall be carried out in accordance with the relevant provisions of BS 594987.

2. Where laying against an exposed edge, the joint shall be treated with an application of bitumen conforming to BS EN 12591 Grade 100/150.

8.7 WATERPROOF UNDERLAY FOR CONCRETE CARRIAGEWAYS

1. Waterproof underlay shall consist of waterproof paper or impermeable plastic sheeting, laid to provide a membrane immediately below the concrete. Overlaps shall be not less than 300 mm and precautions shall be taken to prevent ponding on the membrane.

8.8 REINFORCEMENT OF CONCRETE CARRIAGEWAYS

(i) This Clause assumes that only a single layer of reinforcement is necessary.

1. Reinforcement in concrete carriageways shall be so placed that, after compaction of the concrete, its cover below the finished surface of the slab is 60 mm (± 10 mm) and it terminates 125 mm (± 25 mm) from the edges of the slab and from all pre-formed joints in the concrete.

2. At transverse joints between mats of steel fabric reinforcement, the longitudinal bars shall overlap by at least 35 times the bar diameter, provided that such overlap shall not, in any case, be less than 450 mm. Mats shall be placed so as to maintain the same spacing between edge longitudinal bars as in the body of the mat.

3. Reinforcement shall be positioned above dowel bars and tie bars, irrespective of the tolerances on position.

8.9 LAYING CONCRETE CARRIAGEWAYS

(i) The positioning and detailing of movement joints should be described in the Contract.

1. Placing, compacting and finishing of concrete in carriageways shall be carried out in one layer as rapidly as possible and shall be so arranged that, in any transverse section, the time for completion of mixing the first batch of concrete to completion of compaction of that section shall not exceed 2 hours.

2. Carriageway surfaces shall have a screeded finish, which shall be brushed transversely to produce a lightly brushmarked finish, with a 100 mm arris steel trowel finish to sides and at joints.

3. Surface irregularities shall not exceed 3 mm when checked with a 3 m straight edge.

8.10 LAYING KERBS AND CHANNELS

(i) Details of the concrete bed and haunch should be described in the Contract.

1. Kerbs, edgings, channels and quadrants shall be laid and bedded on a layer of Class M12 mortar, either on the concrete carriageway or on a GEN3 concrete foundation, as described in the Contract. They shall be butt-jointed, except where they are laid on concrete carriageways. They shall be provided with joints coincident with the carriageway movement joints, of width and with filler identical to that used in the carriageway joints. All kerbs shall be backed with GEN3 concrete.

2. For radii of 12 m or less, kerbs and channels of the appropriate radius shall be used.

3. Alignment of kerbs and channels shall not deviate by more than 10 mm from that described in the Contract, with no lipping of visible faces.

8.11 FOUNDATIONS FOR FOOTWAYS

(i) A cross-section of the footway construction should be described in the Contract.

1. Foundations for footways shall consist of Type 1 granular sub-base material, spread evenly and compacted in layers of not more than 100 mm thickness.

2. Compaction to the correct levels shall be carried out using a vibratory roller having a static load of at least 1000 kg per metre width of roll.

8.12 LAYING CONCRETE PAVING FLAGS

(i) Paving flags should not be laid on ground which has been either excavated below normal service depth or not backfilled to Highway Reinstatement Specification.

1. Precast concrete flags of 50 and 63 mm thickness (Types A, B, C or D from BS EN 1339) shall be laid to the required falls on sub-base material, as described in the Contract, and bonded with joints and fully bedded on mortar in accordance with BS 7533-4.

2. Precast concrete paving flags 60, 65 or 70 mm thickness (Types E, F or G from BS EN 1339) shall be laid to the required falls on sub-base material, as described in the Contract, bonded with joints at right angles to the kerb and bedded on sand-bed in accordance with BS 7533-4.

3. Flags shall be cut to fit around surface boxes and other furniture and, on circular work where the radius is 12 m or less, shall be radially cut on both edges to the required lines.

8.13 LAYING PAVING BLOCKS

1. Precast concrete block paving shall be laid in accordance with BS 7533-3.

8.14 TOLERANCES FOR FINISHED CARRIAGEWAY SURFACES

1. Finished surfaces at each stage of road construction shall not vary from the levels described in the Contract by more than the following permissible tolerances:

Surface	Permissible Tolerance (mm)
Formation and sub-base	+ 10, - 30
Base	± 15
Wearing surface or slab surface	± 6

2. The combination of permitted tolerances in the levels of different pavement courses shall not result in a reduction in thickness of the pavement, excluding the sub-base, by more than 15 mm from the specified thickness, nor a reduction in the thickness of the bituminous wearing course by more than 5 mm from that specified, and the maximum allowable irregularity of the wearing surface below a 3 m straight edge shall be 3 mm.

3. Combinations of reverse tolerances shall not result in ponding of surface water.

8.15 FIXING OF GULLIES

1. Gullies shall be bedded and surrounded with GEN3 concrete to the thickness described in the Contract.

2. Frames shall be bedded in epoxy mortar on two courses of Class B engineering brickwork or precast concrete gully cover slabs.

8.16 LINING AND SIGNAGE

(i) Road traffic signs should be provided in accordance with the the Traffic Signs Regulations and General Directions, 2002, published by The Stationery Office.

(ii) Deployment of traffic signs should be as 'Safety at Street Works and Road Works, 2002, published by The Stationery Office.

1. Permanent road markings shall be applied, in accordance with BS 3262-3.

8.17 CONCRETE FOOTWAYS

1. Where in-situ concrete footways are constructed, the final surface shall have a brush finish, with a minimum 60 mm smooth trowelled margin at the edge.

SECTION 9

SEWER RENOVATION

The Notes for Guidance are not part of the Specification.

(i) This Section covers the fundamental requirements of sewer renovation based on the recommendations of the 'Sewerage Rehabilitation Manual', (5th Edition), 2001 published by WRc plc.

(ii) The Contract should describe requirements for:

(a) the interface with the existing operational system to avoid unplanned disruption of services to customers and pollution incidents;

(b) dealing with any unusually aggressive effluents; and

(c) dealing with existing flows to maintain levels of service.

(iii) Attention is directed to BS EN 13566 'Plastics Piping Systems for Renovation of Underground Non-pressure Drainage and Sewerage Networks, Part 1: General, Part 2: Lining with Continuous Pipes, Part 3: Lining with Close Fit Pipes, Part 4: Lining with Cured-in-place Pipes'.

(iv) Reference should be made to Report UM 1464 'Typical Specifications for Trenchless Technology Systems', 1995 published by WRc plc.

9.1 ISOLATION OF FLOWS

1. Flows shall be positively dealt with or isolated from the section being renovated.

(i) Example flow schedules are given in the 'Sewerage Rehabilitation Manual' (5th Edition), 2001 published by WRc plc.

(ii) Where the Contractor will be permitted to operate flow control devices and isolate flows, this should be described in the Contract.

9.2 PREPARATORY SURVEY

1. The sewers to be renovated shall be surveyed to determine the position, size and angle of approach of all laterals, to an accuracy appropriate to the method of reconnection.

2. A survey of the structural condition of the sewers and associated flow control devices shall be carried out.

3. All flow control devices shall be operable prior to the commencement of sewer renovation works.

4. Prior to ordering any materials, the sewer dimensions shall be confirmed from the survey.

(i) Details of the structural condition survey should be described in the Contract.

5. Any sewer pressure main to be renovated shall be surveyed to determine its position, size, material and structural condition, and the location and operability of all fittings that may impede the insertion of the liner.

9.3 PREPARATION OF SEWERS

(i) The Contract should describe where flow control devices may be operated.

1. Sewers to be renovated shall be prepared so that the installation and performance of the renovation system is not impaired. The installation method and preparation shall not affect the stability of the existing sewer.

2. Debris shall be prevented from being carried downstream whilst preparing or working in existing sewers. Such debris is to be removed from the sewer.

3. Intruding laterals shall be finished flush prior to the application of any lining system.

4. Loose or unsound brickwork or masonry in man-entry sewers shall be repaired using Class A engineering bricks and Class M12 mortar, with all joints completely filled before other phases of rehabilitation.

5. All substrate surfaces shall be stated in the Contract.

6. Pressure pipelines shall be thoroughly cleaned to remove all rust, tuberculation, deposits, loose or deteriorated remains of any original coatings and other foreign materials from the inside of the pipe, to produce a smooth surface finish.

7. Valves, bends and other fittings shall be removed, as appropriate, to enable the liner to be correctly installed.

9.4 JOINTING GENERALLY

1. Jointing surfaces and components shall be kept clean and free from extraneous matter until the joints have been made or assembled.

9.5 CONNECTIONS

(i) The period between sewer lining and reconnection of laterals should be described in the Contract.

1. All laterals and branches shall be reconnected following sewer renovation works. Any circumferential lining joints shall not be made at, or within 100 mm of, a connection.

2. Each finished connection shall be made flush with the lining and shall provide a smooth transition.

3. Cured-in-place lateral connection collars ("Top Hats") shall be Class B in accordance with BS EN 13566-4. The rim shall extend for more than 50% of the circumference of the main pipe.

9.6 CHAMBERS

1. In chambers, where applicable, the edges of the lining shall be adequately sealed and the benching made good.

2. Brick chambers shall be repaired with Class B Engineering bricks bedded on Class M12 mortar.

3. Exposed edges or ends of plastics liners shall be mechanically anchored.

9.7 RELEASE OF CURING WATER

(i) Due regard must be paid to the disposal of curing liquors to avoid pollution.

(ii) The Environment Agency, SEPA and NIEA provide advice to industry and the public on their legal responsibilities and good environmental practice through a series of Pollution Prevention Guidance

1. Whenever hot water curing of a lining system is carried out, the water shall not be released until it cools below 40°C.

2. Prior approval from the Client shall be obtained before the release of curing water.

Notes (PPGs), see www.environment-agency.gov.uk.

9.8 ANNULUS GROUTING GENERALLY

(i) Any alternative restrictions on maximum grout pressures should be described in the Contract.

(ii) The grouting materials should be detailed in the Contract.

1. All air, water and contaminated grout are to be expelled from the annulus by the grouting process.

2. The grout injection pressure shall be continuously measured by a suitable gauge and shall not exceed 50 kN/m^2.

3. The quantity of grout injected at each injection point and the maximum pressure at the nozzle at that point shall be recorded.

4. Disposal of grout to live sewers is not permitted.

5. Annulus grouting shall closely follow the placing of any lining segments.

6. The applied grouting pressure shall not exceed the design external pressure for the liner unless the liner is internally pressurised during grouting or other internal support provided.

9.9 INSPECTION AFTER GROUTING

1. The main sewer, branches or laterals shall be inspected immediately after each grouting operation and any excess grout cleaned out.

9.10 LINING THROUGH FLOW CONTROL DEVICES

1. Any flow control devices lined in-situ shall be operated throughout their full travel, several times immediately after lining, in order to ensure free operation.

9.11 INSPECTION AFTER RENOVATION

(i) The format of the information required, and any need for a CCTV survey, should be described in the Contract.

(ii) The number of samples to be removed and the length of the sample should be described in the Contract. Where appropriate, a control sample should be provided, in advance of the renovation commencing, to compare with samples taken at completion.

(iii) The sampling and acceptance of cured-in-place lateral connection collars should be as detailed in the Contract.

1. On completion of the renovation, a survey shall be undertaken and a record of that survey provided to the Client. Where man-entry is not possible, a CCTV survey shall be carried out, including the use of a pan and tilt camera, where laterals are included in the length concerned.

2. Pipe samples shall be taken to check lining performance and to determine the thickness. The lining shall be made good after the samples have been taken. The monitoring of renovation will be used to determine the location of samples. The pipe samples shall be provided to the Client.

3. The method of localised air testing shall be in accordance with Clause 13 of BS EN 1610:1998.

9.12 LINING TEMPLATE

1. For man-entry sewers, a lining template shall be drawn through the sewer prior to ordering the lining to "prove" the suitability of the profile. If necessary, the template shall be adjusted and the proving operation repeated until a suitable profile is established.

2. For non-man-entry sewers, immediately prior to installation a proving pig with solid ends of sufficient rigidity and length to represent the proposed lining system shall be passed down the sewer length to be relined.

9.13 LINING DESIGN

1. The lining design shall be in accordance with the 'Sewerage Rehabilitation Manual' (5th Edition), 2001, published by WRc plc.

9.14 TOLERANCES FOR PREFORMED LININGS

1. The position of the internal face of any unit shall not deviate by more than ± 20 mm from the agreed profile. No section of lining shall have a backfall.

2. Lipping between adjacent units shall not exceed 5 mm.

9.15 CUTTING AND PROTECTION OF LININGS

1. Linings shall be cut in accordance with the manufacturer's instructions.

2. Cut edges of GRP shall be sealed in accordance with WIS 4-34-02.

3. The ends of polyethylene and polypropylene linings shall be mechanically anchored using suitable fixings.

4. At the upstream end of the lining, the original invert shall be broken out and the lining unit laid on a Class M12 mortar bed so that the inverts of the lining and the host pipe are correctly aligned.

9.16 MAN-ENTRY PREFORMED GRP AND GRC UNITS

1. Each lining unit shall be laid to the mean line and level of the existing sewer and be packed, in a minimum of four suitably-spaced positions, with chemically inert spacers of no greater strength than the grout used to fill the annulus.

2. The annulus shall not be less than 10 mm in width.

9.17 CURED-IN-PLACE LINING SYSTEMS

1. All cured-in-place linings shall be designed, manufactured, constructed, tested, inspected and certified in accordance with BS EN 13566-4.

2. Cured-in-place linings for resisting infiltration shall pass the leak tightness test set out in WRc plc Report CP308 'Procedures for Testing the Capability of Cured-in-place Pipe (CIPP) Lining Systems to Resist Infiltration'.

9.18 RENDERING AND LOCAL REPAIRS

(i) Any requirements for rendering and local repairs should be described in the Contract.

1. All bricks intruding into the barrel shall be removed and replaced from inside the sewer.

9.19 PATCH REPAIRS

1. Cured-in-place patch repairs shall be leaktight and shall comply with WIS 4-34-06.

2. The materials used to fabricate the patch shall conform to the requirements of Clause 4.2 of WIS 4-34-06:2010.

3. The patch linings shall be fabricated in accordance with Clause 4.3 of WIS 4-34-06:2010.

4. The materials used for cured-in-place patch repairs for structural enhancement shall incorporate a means of anchorage that will continue to function after all initial shrinkage has occurred.

5. Where a complete pipe length (from joint to joint) requires a repair, the patch shall extend the full length of the pipe (joint to joint) and beyond the pipe joints by a minimum of 100 mm.

6. Where a repair is required within a pipe length, the repair shall extend by a minimum of 100 mm either side of the defect, except where the repair crosses a joint, then Clause 9.19.5 shall apply.

7. Where it is necessary to overlap patches, the overlap shall be a minimum of 150 mm.

8. The Contractor shall specify the method of installation in accordance with Clause 4.3 of WIS 4-34-06:2010.

9.20 STYRENE FUMES

(i) To prevent styrene fumes being forced into laterals and properties, air extraction should be used to draw the fumes out of the sewer from the working manhole. Extractor fans should ventilate the lined sewer from upstream to downstream.

1. The Contractor shall take all precautions to prevent styrene fumes from cured-in-place linings or patch repairs entering the properties via their laterals. A suitable monitor to measure the styrene fumes level shall also be available on Site.

9.21 SPRAY LINING

(i) In the absence of a standard for spray polymeric linings for wastewater applications, the guidance given in BS EN 13566 should be followed.

(ii) The method of testing polymeric spray linings should be as stated in the Contract.

1. Dimensional requirements, material tests and measurements are to be in accordance with BS EN 13566-4.

SECTION 10

WATER MAINS RENOVATION

The Notes for Guidance are not part of the Specification.

(i) This Section covers the fundamental requirements of water mains rehabilitation based on the recommendations of the 'Water Mains Rehabilitation Manual', 1999, published by WRc plc.

(ii) The Contract should describe requirements for:

(a) the interface with the existing operational system to avoid unplanned disruption of services to customers and pollution incidents;

(b) dealing with existing flows to maintain levels of service; and

(c) where valves and hydrants may be operated.

10.1 PREPARATORY SURVEY

1. The water mains to be rehabilitated shall be surveyed to determine their position, size, material and structural condition, and the location and operability of all fittings.

10.2 PREPARATION OF WATER MAINS

1. Water mains to be re-lined shall be prepared internally such that the performance of the lining system is not impaired.

10.3 IN-SITU LININGS

(i) Due regard should be paid to the release of toxic gases and odours from the installation of epoxy lining systems.

1. The measures to be taken to control the release of odours and gases arising from the Works shall be stated in the Contract.

2. The in-situ lining of water mains with epoxy resin shall be in accordance with the, 'Code of Practice: In-situ Resin Linings of Water Mains (IGN 4-02-02) and Operational Requirements: In-situ Resin Lining of Water Mains (WIS 4-02-01) published by Water UK.

3. The in-situ lining of water mains with cement mortar shall be in accordance with the 'In-situ Cement Mortar Lining – Operational Guidelines and Code of Practice', published by WRc plc.

4. The in-situ lining of water mains with polymeric linings shall be in accordance with the 'Code of Practice: In-situ Resin Linings of Water Mains (IGN 4-02-02) and the Operational Requirements: In-situ Resin Lining of Water Mains (WIS 4-02-01), published by Water UK.

10.4 SLIPLINING AND REPLACEMENT MOLING

(i) The period between pipe lining and reconnection of services should be described in the Contract.

1. The use of sliplining and moling techniques shall be in accordance with 'Sliplining and Replacement Moling of Water Mains – Operational Guidelines' published by WRc plc.

2. The annular void shall be grouted unless otherwise stated in the Contract.

10.5 CONNECTIONS

1. Service connections shall be disconnected and reconnected, as required, for the work.

10.6 LINING THROUGH VALVES

1. Any valves lined in-situ shall be operated throughout their full travel, several times immediately after lining, in order to ensure free operation. There shall be no in-situ lining through valves of pipelines of non-man-entry sizes.

10.7 INSPECTION AFTER IN-SITU RE-LINING

(i) The format of the information required and CCTV survey should be described in the Contract.

(ii) The number of samples to be removed and the length of the sample should be described in the Contract. Where appropriate, a control sample should be provided, in advance of the renovation commencing, to compare with samples taken at completion.

1. On completion of the rehabilitation, a survey shall be undertaken and a record of that survey shall be supplied to the Client.

2. Pipe samples shall be taken and provided to the Client in order to check lining performance and to determine thickness. The lining shall be made good after the samples have been taken. The monitoring of rehabilitation will be used to determine the location of samples.

10.8 BRINGING REHABILITATED WATER MAINS INTO SERVICE

1. Before returning any rehabilitated water main into service, the main shall be cleaned and disinfected.

SECTION 11

TUNNELLING AND SHAFT SINKING WORKS

The Notes for Guidance are not part of the Specification.

11.1 HEADINGS, TUNNELS AND SHAFTS

(i) For further guidance, see British Tunnelling Society and Institution of Civil Engineers 'Specification for Tunnelling'.

(ii) Where appropriate, particularly in areas of previous mining activity, probing ahead requirements should be described in the Contract.

(iii) This Section, as well as dealing with tunnelling and shaft sinking, provides for thrust boring.

(iv) The need for any ground investigation ahead of the face of any tunnel, heading or shaft should be described in the Contract.

1. Any restrictions on vehicle movements and working hours shall be as stated in the Contract.

2. On Site storage capacity required for excavated materials shall be stated in the Contract.

3. Excavations for headings and tunnels shall be adequately supported at all times. The faces of all headings and tunnels, other than those in solid rock, shall be fully supported whenever excavation has been stopped for a period exceeding 12 hours.

4. All working shafts shall be supplied with effective means of communication with the Works below ground. A fail-safe back-up means of communication shall also be provided.

11.2 SHAFTS

(i) Any special requirements relating to protection should be described in the Contract.

(ii) Special Clauses will be required for other types of shaft sinking.

1. A safe means of access and egress shall be provided in compliance with BS 6164. All shafts shall be provided with a primary and secondary means of access and egress. A ladder access shall be provided in addition to any mechanical means.

2. Segments used in shafts shall be so erected as to break vertical joints, except in any rings required to be broken out.

3. Where shafts are constructed by underpinning and lined with segments, they shall be grouted up at least once per shift.

4. Where segmental shafts are sunk by caisson methods, an annulus shall be maintained by a lubricant which shall be replaced with a grout on completion.

5. After any primary lining is complete and before any openings are made at or near the foot, the excavation for the base of the shaft shall be taken out and the base concreted.

11.3 OPENINGS IN SHAFTS AND TUNNELS

(i) The Contract should describe any limitations which the design of the tunnel or shaft will impose on temporary openings.

1. Drawings shall be provided showing proposals for forming openings in shafts and tunnels. These Drawings shall include details of temporary supports to the lining and to the ground.

2. Openings in shafts and tunnels shall only be made after the segments have been grouted.

11.4 SEGMENTAL SHAFT AND TUNNEL LININGS

1. Segments shall be erected and assembled in accordance with the manufacturer's instructions and the Contractor's method statement. The lining shall be built as soon as possible after the ground has been cut.

11.5 UNBOLTED CONCRETE TUNNEL SEGMENTS

(i) Requirements for any circumferential pre-stress should be described in the Contract.

1. The shape of unbolted concrete segmental tunnels shall be maintained within tolerance after erection, until the segments have been stabilised by grout or other means.

2. Where a circumferential pre-stress is applied, the force shall be such that the whole of the concrete lining is expanded tight against the surrounding ground. A joint lubricating compound shall be applied to all wedge faces of each segment to be expanded.

3. Where wedge block segments are specified, the excavated profile shall be lubricated to reduce skin friction.

4. Where key segments are shorter than other segments comprising the ring, the pockets formed shall be filled with concrete to the grade specified in the Contract.

11.6 BOLTED CONCRETE SEGMENTAL LININGS

(i) Any requirement relating to rolling of segments should be described in the Contract.

1. Segmental joints to bolted concrete tunnel and shaft linings shall be trued and longitudinal joint bolts tightened before the final tightening of the circumferential joint bolts connecting the ring to the adjacent ring.

2. Packings shall be inserted in the joints of the lining at the time of erection, as required, to maintain correct shape, line and level, and shall not exceed 6 mm or half of the sealing capacity of the gasket.

3. Segments shall be fixed in accordance with manufacturer's recommendations. Where grummets are to be used, two grummets shall be threaded on each bolt to be grummetted at the time any bolted segment is erected. One grummet shall be placed under the washer at the head of the bolt and the other under the washer at the nut.

11.7 GROUTING OF SEGMENTS

(i) Any requirements for high pressure grouting should be described in the Contract.

(ii) Any locations where it is required to grout more frequently than once per shift should be described in the Contract.

(iii) When selecting grout, due regard should be given to its specification to ensure that local aquifers are not polluted.

(iv) A pressure gauge should be fitted onto the grouting nozzle to avoid ground heave, overstress or distortion.

1. Segmental shaft and tunnel linings shall be grouted by forcing the required grout through the grouting holes in the segments, so that all interstices around the outside of the segments are filled. Adequate venting shall be provided to ensure that air is not trapped.

2. Temporary hardwood plugs shall be inserted into grout holes after grouting; these shall be replaced by permanent plugs of material similar to that of the segments being grouted when it is evident to the Client that grouting has been effective.

3. Grout pipes shall be provided in head walls or ring walls, and any void remaining after concreting shall be filled with the required grout.

4. Where primary grouting does not completely fill all cavities, secondary grouting shall be carried out.

5. Grouting shall closely follow the erection of rings, be undertaken every four rings, and at least twice per shift.

11.8 CAULKING

(i) Material to be used for caulking should be described in the Contract.

(ii) The time when caulking is carried out may depend, for example, upon ground conditions and availability of the working area.

1. Caulking of segment joints in tunnels and shafts shall be carried out as late as practicable within the construction programme; the grooves shall be raked out and cleaned immediately before caulking.

2. Caulking of circumferential and longitudinal joints shall be bonded to form a homogeneous and continuous mass, consolidated to fill the recess up to the inner surface of the segment or to the depths described in the Contract.

11.9 POINTING OF JOINTS

1. Where joints between segments are required to be pointed, they shall be raked out, cleaned, filled with a proprietary non-shrink cement mortar and pointed flush.

11.10 SECONDARY LININGS TO SEGMENTS

(i) The specification of the lining and the surface finish should be stated in the Contract.

1. Before construction of a secondary internal lining to a segmental tunnel is commenced, it shall be cleaned out, caulked and proven to be watertight.

2. The minimum thickness of lining over the ribs of the segments shall be agreed in advance with the Client, unless otherwise stated in the Contract.

11.11 WATERTIGHTNESS OF SHAFTS AND TUNNELS

(i) See also Clause 7.8.

1. Shafts and tunnels shall be substantially watertight, with no identifiable flow of water penetrating either the primary or secondary lining.

11.12 CONTROL OF GROUNDWATER

1. Working methods and systems shall be designed to control groundwater and, where necessary, to remove water from the tunnel. Removal of groundwater shall not cause damage to the Works nor to the property of third parties.

11.13 PIPE JACKING

(i) Advice on the design and construction of pipejacks is given in 'A Guide to Best Practice for the Installation of Pipejacks and Microtunnels', published by the Pipe Jacking Association.

(ii) Collars fabricated from weldable structural steel may be susceptible to corrosion from the ground, groundwater or the effluent carried. If corrosion can be expected, the design of the joint should provide for a secondary sealing gasket to be applied on Site.

(iii) The material required to seal the joints should be described in the Contract.

1. Excavation for pipe jacking shall be undertaken from within a shield equipped with steering jacks for adjusting the alignment. Face boards shall be available for boarding up the exposed excavation.

2. The jacking load applied to the pipeline shall be limited so that damage to the pipes is avoided and an intermediate jacking station shall be provided, where necessary, to achieve this.

3. The jacking load shall be transferred to the pipes through a thrust ring, which shall be sufficiently rigid to ensure even distribution of the load.

4. The pipe manufacturer's permitted draw or angular deflection shall not be exceeded at any individual joint.

5. Up-to-date records shall be maintained of jacking loads and line and level measurements. A graphical relationship between the jacking force and the distance moved shall be produced to ensure that the necessary measures are taken to avoid exceeding the maximum permitted jacking forces.

6. All lifting holes and grouting holes shall be sealed.

7. Joint packing material designed to distribute the jacking load evenly shall be inserted at, and between, the pipe ends, and at any intermediate jacking stations.

8. Before work may start on any thrust pit, it shall be demonstrated that the design will withstand the maximum jacking force of which the jacks are capable.

9. Before a particular pipe jack may commence, evidence shall be provided that the pipes necessary for completing the jack are on Site or in storage elsewhere. The envisaged number of intermediate jacking stations shall be agreed in advance of commencement. At least one intermediate jacking station assembly shall be available at the start of the pipe jack, unless otherwise agreed.

10. Pipes, including lead pipes, which have been jacked through a pipe jack shall not be used elsewhere on the Works. Cut pipes shall not be jacked.

11. Pipe jacking shall be undertaken in a continuous operation and the jacking load shall be limited so that installed joints are prevented from opening when the jacking loads are removed.

12. On completion of the drive, intermediate jacking stations shall be left fully closed. All jacks, props, thrust rings and packing shall be removed, the ends of the pipes cleaned, a new packing ring glued to the receiving face and the joint jacked partly closed. An "O" ring seal shall then be inserted into the sliding joint and the joint jacked fully closed. The order of closing the stations shall be from the tunnelling shield working back.

13. The annular space between the sides of the excavated tunnel and the jacking pipes shall be constantly filled with an approved lubricant at a pressure that will support the adjacent excavation. Daily records of the quantity of lubricant used for each length of pipe thrust and the point at which the lubricant was injected shall be submitted to the Client. The lubricant shall be thoroughly mixed prior to pumping or placing. On completion of the pipe jack, the annulus shall be filled by displacing the lubricant with grout.

11.14 MICROTUNNELLING

1. The microtunnelling machine shall be selected with regard to the ground conditions, length of drive and other relevant factors.

2. Microtunnelling shall comply, generally, with the provisions of Clause 11.13.

11.15 VENTILATION OF TUNNELS AND SHAFTS

(i) See also the 'Construction (Health, Safety and Welfare) Regulations', 1996, the 'Construction (Health, Safety and Welfare) Regulations' (Northern Ireland) 1996, BS 6164 and HSE Guidance Note EH40.

1. Sufficient ventilation throughout the tunnels and shafts shall be provided to ensure that a safe working environment is maintained at all times.

11.16 WORK IN COMPRESSED AIR

(i) The 'Work in Compressed Air Regulations', 1996, (and the 'Work in Compressed Air Regulations (Northern Ireland)', 2004) and the accompanying Guidance Notes should be referred to when carrying out work in compressed air.

(ii) The CIRIA Report R44 gives recommendations as to standards of good practice for work in compressed air.

(iii) The appropriate provision of BS 6164 also applies.

1. Full details shall be submitted to the Client of the type, capacity and arrangement of the compressed air plant, ancillary equipment and medical facilities it is proposed to install.

2. All standby plant shall be tested weekly by using it to supply air to the Works under normal working conditions.

3. After installation, but before work in compressed air is commenced, all compressed air plant shall be given a 24-hour continuous running test at the normal operating pressure.

11.17 RECORDING OF INFORMATION

1. Records shall be kept of the line, level and the diameter, measured horizontally and vertically, of any tunnel lining. Similar records shall be kept for shafts and for pipe jacking.

11.18 TOLERANCES FOR SHAFTS, TUNNELS AND PIPE JACKS

(i) Where a tunnel is to have a secondary lining, it may be necessary to specify permissible deviations for the primary lining, having regard to the nature of the work.

1. The position of the internal face of any shaft or tunnel shall not deviate from that described in the Contract by more than the following permissible tolerances:

(ii) Any requirements for more stringent tolerances should be described in the Contract.

(iii) Where a shaft or tunnel is to be constructed in ground which is variable or unstable, it may be appropriate for a larger tolerance for line and level to be described in the Contract.

Work Category	Dimension or Alignment	Permissible Tolerances
Pipe jacking/ Microtunnelling	Line	± 50 mm
	Level	± 35 mm
Shafts and chambers	Vertically	1 in 300
Shafts and tunnels	Finished diameter	1% but not exceeding ± 50 mm
Tunnels without secondary lining	Line	± 35 mm
	Level	± 35 mm
Tunnels with secondary lining	Line	± 35 mm
	Level	± 35 mm
Shaft, tunnel and sewer lining segments	Maximum lipping between edges of adjacent segments	10 mm

2. The difference between maximum and minimum measured diameters in any one segmental ring shall not exceed 2% of the diameter of the ring.

3. No combination of tolerances resulting in a reverse fall will be permitted where the invert of the tunnel is required to convey sewage by gravity flow.

APPENDIX I

INCORPORATION OF SPECIFICATION INTO CONTRACTS

It is not necessary to bind copies of the 'Civil Engineering Specification for the Water Industry (7th Edition)', into the documents prepared for Tenderers or into any formally sealed Contract.

Supplementary Clauses should be included in the Contract and numbered in two separate groups as follows:

(a) Specification requirements which are related to an existing Clause should be numbered as additional Sub-clauses (e.g., requirements for special flanges other than to BS EN 1092 should be numbered 2.36.10.)

(b) New Clauses unrelated to existing Clauses should be numbered to follow the last Clause of the appropriate Section.

The supplementary Clauses for each Water Company may be found at www.wrcplc.co.uk/ceswi.

The Specification should be incorporated, by reference, by including the following provisions, in Tender Documents, immediately preceding the Supplementary Specification Clauses as follows:

SPECIFICATION

1. The Specification referred to in the Tender shall be the 'Civil Engineering Specification for the Water Industry (7th Edition)', published by UK Water Industry Research Ltd in March 2011, augmented by the Supplementary Clauses.
2. In so far as any Supplementary Clause may conflict, or be inconsistent, with any provision of the 'Civil Engineering Specification for the Water Industry (7th Edition)', the Supplementary Clause shall always prevail.

APPENDIX II

LIST OF STANDARDS TO WHICH REFERENCE IS MADE IN THE DOCUMENT

STANDARD TYPES

BS	British Standard
BS EN	European Standard adopted as a British Standard
BS EN ISO	International Standard adopted as a British Standard
CP	Code of Practice
DD	Drafts for Development
ISO	International Organisation for Standardisation
PD	Published Documents
prEN	Draft European Standard
PAS	Product Assessment Specification

STD TYPE	STD NO	TITLE
BS	4	Structural Steel Sections
		Part 1: Hot-rolled Sections
PAS	27	Unplasticised Poly(Vinyl Chloride) Alloy (PVC-A) Pipes and Bends for Water Under Pressure
BS EN	33	Pedestal W.C. Pans With Close-coupled Flushing Cistern. Connecting Dimensions
BS EN	37	Pedestal W.C. Pans With Independent Water Supply. Connecting Dimensions
BS ISO	37	Rubber, Vulcanized or Thermoplastic. Determination of Tensile Stress-strain Properties
BS ISO	48	Rubber, Vulcanized or Thermoplastic. Determination of Hardness (Hardness Between 10 IRHD and 100 IRHD)
BS	65	Specification for Vitrified Clay Pipes, Fittings and Ducts, also Flexible Mechanical Joints for Use Solely With Surface Water Pipes And Fittings
BS EN	111	Wall-hung Hand Rinse Basins. Connecting Dimensions
BS EN	124	Gully Tops and Manhole Tops for Vehicular and Pedestrian Areas. Design Requirements, Type Testing, Marking, Quality Control
DD	140	Wall Ties
		Part 2: Recommendations for Design of Wall Ties
BS EN ISO	150	Raw, Refined and Boiled Linseed Oil for Paints and Varnishes. Specifications and Methods of Test
BS EN	197	Cement
		Part 1: Composition, Specifications and Conformity Criteria for Common Cements Part 4: Composition, Specifications and Conformity Criteria for Low Early Strength Blastfurnace Cements
BS EN	200	Sanitary Tapware. Single Taps and Combination Taps for Water Supply Systems of Type 1 and Type 2. General Technical Specification
BS EN	206	Concrete
		Part 1: Specification, Performance, Production and Conformity
DD	256	Assessment of the Potential for Metallic Materials to Affect Adversely the Quality of Water Intended for Human Consumption. Specification
BS EN	274	Waste Fittings for Sanitary Appliances
		Part 1: Requirements
BS EN	287	Approval Testing of Welders for Fusion Welding
		Part 1: Steels
BS EN	295	Vitrified Clay Pipes and Fittings and Pipe Joints for Drains and Sewers
		Part 1: Requirements
		Part 4: Special Fittings, Adaptors and Compatible Accessories
		Part 5: Requirements for Perforated Vitrified Clay Pipes and Fittings
		Part 7: Requirements for Vitrified Clay Pipes and Joints for Pipe Jacking
BS EN	301	Adhesives, Phenolic and Aminoplastic, for Loading-Bearing Timber Structures, Classification and Performance Requirements
BS EN	338	Structural Timber. Strength Classes
BS	410	Test Sieves. Technical Requirements and Testing
		Part 1: Test Sieves of Metal Wire Cloth
		Part 2: Test Sieves of Perforated Metal Plate
BS EN	413	Masonry Cement
		Part 1: Composition, Specifications and Conformity Criteria
BS	416	Discharge and Ventilating Pipes and Fittings, Sand-Cast or Spun in Cast Iron
		Part 1: Specification for Spigot and Socket Systems
BS	417	Specification for Galvanised Low Carbon Steel Cisterns, Cistern Lids, Tanks and Cylinders

STD TYPE	STD NO	TITLE
		Part 2: Metric Units
BS	434	Bitumen Road Emulsions (Anionic and Cationic)
		Part 1: Specification for Anionic Bitumen Road Emulsions
BS EN	450	Fly Ash for Concrete
		Part 1: Definitions, Specifications and Conformity Criteria
BS	459	Specification for Matchboarded Wooden Door Leaves for External Use
BS EN	459	Building Lime. Definitions, Specifications and Conformity Criteria
		Part 1: Definitions, Specifications and Conformity Criteria
BS	460	Specification for Cast Iron Rainwater Goods
BS	466	Specification for Power Driven Overhead Travelling Cranes, Semi-goliath and Goliath Cranes for General Use
BS EN	485	Aluminium and Aluminium Alloys. Sheet, Strip and Plate
		Part 1: Technical Conditions for Inspection and Delivery
BS EN	490	Concrete Roofing Tiles and Fittings. Product Specifications
BS	493	Specification for Air Bricks and Gratings for Wall Ventilation
BS EN	520	Gypsum Plasterboards. Definitions, Requirements and Test Methods
BS	544	Specification for Linseed Oil Putty for Use in Wooden Frames
BS EN	545	Ductile Iron Pipes, Fittings, Accessories and their Joints for Water Pipelines. Requirements and Test Methods
BS EN	571	Non-destructive Testing. Penetrant Testing
		Part 1: General Principles
BS EN	573	Aluminium and Aluminium Alloys
		Part 3: Chemical Composition and Form of Wrought Products. Chemical Composition and Form of Products
BS EN	593	Industrial Valves. Metallic Butterfly Valves
BS EN	598	Ductile Iron Pipes, Fittings, Accessories and their Joints for Sewerage Applications. Requirements and Test Methods
BS EN	607	Eaves Gutters and Fittings Made of PVC-U. Definitions, Requirements and Testing
BS EN	612	Eaves Gutters and Rainwater Down-Pipes of Metal Sheet. Definitions, Classifications and Requirements
BS EN	622	Fibreboards. Specifications
		Part 1: General Requirements
		Part 2: Requirements for Hardboards
		Part 3: Requirements for Medium Boards
		Part 4: Requirements for Softboards
		Part 5: Requirements for Dry Process Boards (MDF)
BS EN	634	Cement-bonded Particle Boards. Specification
		Part 1: General Requirements
		Part 2: Requirements for OPC Bonded Particleboards for Use in Dry, Humid and Exterior Conditions
BS EN	636	Plywood. Specifications
BS EN	639	Common Requirements for Concrete Pressure Pipes Including Joints and Fittings
BS EN	642	Prestressed Concrete Pressure Pipes, Cylinder and Non-cylinder, Including Joints, Fittings and Specific Requirement for Prestressing Steel for Pipes
BS	644	Wood Windows
		Part 1: Specification for Factory Assembled Windows of Various Types
		Part 3: Wood Double Hung Sash and Case Windows (Scottish Type)
BS EN	649	Resilient Floor Coverings. Homogeneous and Heterogeneous Polyvinyl Chloride Floor Coverings. Specification
BS EN	654	Resilient Floor Coverings. Semi-Flexible Polyvinyl Chloride Tiles. Specification
		Part 2: Metric Units
BS EN	681	Elastomeric Seals. Material Requirements for Pipe Joint Seals Used in Water and Drainage Applications *(Supplement to BS2494)*
		Part 1: Vulcanized Rubber
		Part 2: Thermoplastic Elastomers
BS	743	Specification for Materials for Damp-Proof Courses
BS	750	Specification for Underground Fire Hydrants and Surface Box Frames and Covers
BS EN	752	Drain and Sewer Systems Outside Buildings
BS EN	754	Aluminium and Aluminium Alloys. Cold Drawn Rod/Bar and Tube
		Part 1: Technical Conditions for Inspection and Delivery
BS EN	755	Aluminium and Aluminium Alloys. Extruded Rod/Bar, Tube and Profiles

STD TYPE	STD NO	TITLE
		Part 1: Technical Conditions for Inspection and Delivery
BS EN	756	Welding Consumables. Wire Electrodes and Wire-Flux Combinations for Submerged Arc Welding of Non Alloy and Fine Grain Steels. Classification
BS EN	771	Specification for Masonry Units
		Part 1: Clay Masonry Units
		Part 2: Calcium Silicate Masonry Units
		Part 3: Aggregate Concrete Masonry Units (Dense and Light-weight Aggregates)
		Part 5: Manufactured Stone Masonry Units
		Part 1: Determination of Compressive Strength
		Part 2: Determination of Percentage Area of Voids in Aggregate Concrete Masonry Units (by Paper Indentation)
		Part 3: Determination of Net Volume and Percentage of Voids of Clay Masonry Units by Hydrostatic Weighing
		Part 4: Determination of Real and Bulk Density and of Total and Open Porosity for Natural Stone Masonry Units
BS EN	805	Water Supply- Requirements for Systems and Components Outside Buildings
BS EN	806	Specifications for Installations Inside Buildings Conveying Water for Human Consumption
		Part 2: Design
		Part 4: Installation
BS	812	Testing Aggregates
		Part 105: Method for Determination of Particle Shape
		Part 111: Method for Determination of Ten Percent Fines Value (TFV)
		Part 124: Method for Determination of Frost-heave
BS EN	818	Short Link Chain for Lifting Purposes. Safety
		Part 1: General Conditions of Acceptance
		Part 3: Medium Tolerance Chain for Chain Slings. Grade 4
BS EN	845	Specification for Ancillary Components for Masonry
		Part 1: Ties, Tension Straps, Hangers and Brackets
		Part 2: Lintels
		Part 3: Bed Joint Reinforcement of Steel Meshwork
BS EN ISO	845	Cellular Plastics and Rubbers. Determination of Apparent (Bulk) Density
BS EN	877	Cast Iron Pipes and Fittings, their Joints and Accessories for the Evacuation of Water from Buildings. Requirements, Test Methods and Quality Assurance
BS EN	912	Timber Fasteners. Specifications for Connections for Timber
BS EN	927	Paints and Varnishes. Coating Materials and Coating Systems for Exterior Wood
		Part 1: Classification and Specification
BS EN	933	Tests for Geometrical Properties of Aggregates – Determination of Particle Size Distribution
		Part 1: Sieving Method
		Part 3: Flakiness Index
BS EN	934	Admixtures for Concrete, Mortar and Grout
		Part 2: Concrete Admixtures. Definitions, Requirements, Conformity, Marking and Labelling
		Part 3: Admixtures for Masonry Mortar. Definitions, Requirements, Conformity and Marking and Labelling
BS EN	942	Timber in Joinery
BS	952	Glass for Glazing
		Part 1: Classification
		Part 2: Terminology for Work on Glass
BS EN	969	Specification for Ductile Iron Pipes, Fittings, Accessories and their Joints for Gas Pipelines. Requirements and Test Methods
BS EN	997	WC Pans and WC Suites with Integral Trap
BS EN	998	Specification for Mortar for Masonry
		Part 1: Rendering and Plastering Mortar
		Part 2: Masonry Mortar
BS EN	1008	Mixing Water for Concrete. Specification for Sampling, Testing and Assessing the Suitability of Water, Including Water Recovered from Processes in the Concrete Industry, as Mixing Water for Concrete
BS	1010	Specification for Draw-Off Taps and Stopvalves for Water Services (Screw-down Pattern)
		Part 2: Draw-Off Taps and Above-ground Stopvalves
BS EN	1011	Welding. Recommendations for Welding of Metallic Materials
		Part 1: General Guidance for Arc Welding

STD TYPE	STD NO	TITLE
		Part 2: Arc Welding of Ferritic Steels
		Part 3: Arc Welding of Stainless Steels
BS	1052	Specification for Mild Steel Wire for General Engineering Purposes
BS EN	1057	Copper and Copper Alloys. Seamless, Round Copper Tubes for Water and Gas in Sanitary and Heating Applications
BS	1070	Specification for Black Paint (Tar-based)
BS EN	1074	Valves for Water Supply. Fitness for Purpose Requirements and Appropriate Verification Tests
		Part 1: General Requirements
		Part 2: Isolating Valves
		Part 3: Check Valves
		Part 4: Air Valves
		Part 5: Control Valves
BS	1088	Marine Plywood
		Part 1: Requirements
BS EN	1092	Flanges and their Joints - Circular Flanges for Pipes, Valves, Fittings and Accessories
		Part 1: Steel Flanges
		Part 2: Cast Iron Flanges
BS EN	1097	Tests for Mechanical and Physical Properties of Aggregates
		Part 6: Determination of Particle Density and Water Absorption
BS	1125	Specification for WC Flushing Cisterns (Including Dual Flush Cisterns and Flush Pipes)
BS	1161	Specification for Aluminium Alloy Sections for Structural Purposes
BS EN	1171	Industrial Valves. Cast Iron Gate Valves
BS	1186	Timber for and Workmanship in Joinery
		Part 2: Specification for Workmanship
		Part 3: Specification for Wood Trim and its Fixing
BS	1188	Specification for Ceramic Wash Basins and Pedestals
BS	1196	Specification for Clayware Field Drain Pipes and Junctions
BS	1202	Specification for Nails
		Part 1: Steel Nails
		Part 2: Copper Nails
		Part 3: Aluminium Nails
BS	1206	Specification for Fireclay Sinks: Dimensions and Workmanship
BS	1212	Float Operated Valves
		Part 2: Specification for Diaphragm Type Float Operated Valves (Copper Alloy Body, Excluding Floats)
		Part 3: Specification for Diaphragm Type Float Operated Valves (Plastics Bodied) for Cold Water Services Only (Excluding Floats)
BS	1217	Cast Stone. Specification
BS	1245	Specification for Metal Door Frames (Steel)
BS	1254	Specification for WC Seats (Plastics)
BS EN	1254	Copper and Copper Alloys. Plumbing Fittings.
		Part 3: Fittings with Compression Ends for Use with Plastics Pipes
BS EN	1295	Structural Design of Buried Pipelines Under Various Conditions of Loading
		Part 1: General Requirements
BS	1297	Specification for Tongued and Grooved Softwood Flooring
BS EN	1304	Clay Roofing Tiles and Fittings. Product Definitions and Specifications
BS EN	1313	Round and Sawn Timber - Permitted Deviations and Preferred Sizes
		Part 1: Softwood Sawn Timber
BS	1329	Specification for Metal Hand Rinse Basins
		Part 1: Plastics Piping Systems for Soil and Waste Discharge (Low and High Temperature) within the Building Structure. Unplasticized Poly (Vinyl Chloride) (PVC-U). Specifications for Pipes, Fittings and the System
BS	1336	Specification for Knotting
BS EN	1338	Concrete Paving Blocks. Requirements and Test Methods
BS EN	1339	Concrete Paving Flags. Requirements and Test Methods
BS EN	1340	Concrete Kerb Units. Requirements and Test Methods
BS EN	1342	Setts of Natural Stone for External Paving. Requirements and Test Methods
BS EN	1343	Kerbs of Natural Stone for External Paving. Requirements and Test Methods

STD TYPE	STD NO	TITLE
BS EN ISO	1346	Fibre Ropes. Polypropylene Split Film. Monofilament and Multifilament (PP2) and Polypropylene High Tenacity (PP3). 3-, 4- & 8-Strand Ropes
BS EN	1363	Fire Resistance Tests
BS	1377	Methods of Test for Soils for Civil Engineering Purposes
		Part 2: Classification Tests
		Part 4: Compaction-Related Tests
BS	1387	Specification for Screwed and Socketed Steel Tubes and Tubulars and Plain End Steel Tubes Suitable for Welding or for Screwing to BS 21 Pipe Threads
BS EN	1401	Plastics Piping Systems for Non-pressure Underground Drainage and Sewerage. Unplasticised Poly (Vinyl Chloride) (PVC-U)
		Part 1: Specifications for Pipes, Fittings and the System
BS	1438	Specification for Media for Biological Percolating Filters
BS EN ISO	1451	Plastics Piping Systems for Soil and Waste Discharge (Low and High Temperature) Within the Building Structure. Polypropylene (PP)
		Part 1: Specifications for Pipes, Fittings and the System
BS EN	1452	Plastics Piping Systems for Water Supply. Unplasticised Poly(Vinyl Chloride) (PVC-U)
		Part 2: Pipes
		Part 3: Fittings
BS EN	1455	Plastics Piping Systems for Soil and Waste (Low and High Temperature) Within the Building Structure. Acrylonitrile-Butadiene-Styrene (ABS)
		Part 1: Specifications for Pipes, Fittings and the System
BS EN ISO	1461	Hot Dip Galvanised Coatings on Fabricated Iron and Steel Articles. Specifications and Test Methods
BS EN	1462	Brackets for Eaves Gutters. Requirements and Testing
BS	1474	Specification for Wrought Aluminium and Aluminium Alloys for General Engineering Purposes: Bars, Extruded Round Tubes and Sections
BS	1494	Specification for Fixing Accessories for Building Purposes
		Part 1: Fixings for Sheet, Roof and Wall Coverings
BS EN	1504	Products and Systems for the Protection and Repair of Concrete Structures. Definitions, Requirements, Quality Control and Evaluation of Conformity
		Part 10: Site Application of Products and Quality Control of the Works
BS EN	1519	Plastics Piping Systems for Soil and Waste Discharge (Low and High Temperature) Within the Building Structure. Polyethylene (PE)
		Part 1: Specifications for Pipes, Fittings and the System
BS	1521	Specification for Waterproof Building Papers
BS EN	1559	Founding. Technical Conditions of Delivery
		Part 1: General
		Part 4: Additional Requirements for Aluminium Alloy Castings
BS EN	1563	Founding. Spheroidal Graphite Cast Iron
BS EN	1565	Plastics Piping Systems for Soil and Waste Discharge (Low and High Temperature) Within the Building Structure. Styrene Copolymer Blends (SAN + PVC)
		Part 1: Specifications for Pipes, Fittings and the System
BS EN	1566	Plastics Piping Systems for Soil and Waste Discharge (Low and High Temperature) Within the Building Structure. Chlorinated Poly(Vinyl Chloride) (PVC-C)
		Part 1: Specification for Pipes, Fittings and the System
BS EN	1600	Welding Consumables. Covered Electrodes for Manual Metal Arc Welding of Stainless and Heat Resisting Steels. Classification
BS EN	1610	Construction and Testing of Drains and Sewers
BS EN	1668	Welding Consumables. Rods, Wires and Deposits for Tungsten Inert Gas Welding of Non Alloy and Fine Grain Steels. Classification
BS EN	1676	Aluminium and Aluminium Alloys. Alloyed Ingots for Remelting. Specifications
BS	1710	Specification for Identification of Pipelines and Services
BS EN	1714	Non Destructive Testing of Welded Joints. Ultrasonic Testing of Welded Joints
BS	1722	Fences
		Part 1: Specification for Chain Link Fences
		Part 2: Specification for Rectangular Wire Mesh and Hexagonal Wire Netting Fences
		Part 4: Specification for Cleft Chestnut Pale Fences
		Part 5: Specification for Close Boarded Fences
		Part 7: Specification for Wooden Post and Rail Fences
		Part 8: Specification for Mild Steel (Low Carbon Steel) Continuous Bar Fences
		Part 9: Specification for Mild Steel (Low Carbon Steel) Fences with Round or Square Verticals and Flat Horizontals

STD TYPE	STD NO	TITLE
		Part 10: Specification for Anti-Intruder Fences in Chain Link and Welded Mesh
		Part 11: Specification for Woven Wood and Lap Boarded Panel Fences
		Part 12: Specification for Steel Palisade Fences
		Part 14: Specification for Open Mesh Steel Panel Fences
BS EN	1796	Plastics Piping Systems for Water Supply With or Without Pressure. Glass-reinforced Thermosetting Plastics (GRP) Based on Unsaturated Polyester Resin (UP)
BS ISO	1817	Rubber, Vulcanized. Determination of the Effect of Liquids
BS	1876	Specification for Automatic Flushing Cisterns for Urinals
BS EN	1912	Structural Timber. Strength Classes. Assignment of Visual Grades and Species
BS EN	1916	Concrete Pipes and Fittings, Unreinforced, Steel Fibre and Reinforced
BS EN	1917	Concrete Manholes and Inspection Chambers, Unreinforced, Steel Fibre and Reinforced
BS	1968	Specification for Floats for Ballvalves (Copper)
BS EN	1991-1	Actions on Structures. General Actions
		Part 1: Densities, Self-weight, Imposed Loads for Buildings
BS EN	1992	Design of Concrete Structures
		Part 3: Liquid Retaining and Containing Structures
BS EN	1993	Design of Steel Structures
		Part 2: Steel Bridges
BS EN	1993-1	Design of Steel Structures. General rules
		Part 3: Supplementary Rules for Cold-formed Members and Sheeting
BS EN	1995-1	Design of Timber Structures. General
		Part 1: Common Rules and Rules for Buildings
BS EN	1996	Design of Masonry Structures
		Part 2: Design Considerations, Selection of Materials and Execution of Masonry
BS EN	1996-1	Design of masonry structures
		Part 1: General Rules for Reinforced and Unreinforced Masonry Structures
BS EN ISO	2081	Metallic and Other Inorganic Coatings. Electroplated Coatings of Zinc With Supplementary Treatments on Iron or Steel
BS	2456	Specification for Floats (Plastics) for Float Operated Valves for Cold Water Services
BS	2494	Specification for Elastomeric Seals for Joints in Pipework and Pipelines (*Supplemented by BS EN 681-1)*
BS EN ISO	2560	Welding Consumables. Covered Electrodes for Manual Metal Arc Welding of Non-alloy and Fine Grain Steels. Classification
BS ISO	2781	Rubber Vulcanized or Thermoplastic. Determination of Density
BS	2782	Method of Testing Plastics. Glass Reinforced Plastics. Determination of Flexural Properties
		Part 10: Three Point Method
BS	2879	Specification for Draining Taps (Screw-down Pattern)
BS	2971	Specification for Class II Arc Welding of Carbon Steel Pipework for Carrying Fluids
BS	3251	Specification for Indicator Plates for Fire Hydrants and Emergency Water Supplies
BS	3262	Hot-applied Thermoplastic Road Marking Materials
		Part 3: Specification for Application of Material to Road Surfaces
BS	3382	Specification for Electroplated Coatings on Threaded Components
		Part 2: Zinc on Steel Components
BS	3416	Specification for Bitumen-based Coatings for Cold Application, Suitable for Use in Contact with Potable Water
BS	3470	Specification for Field Gates and Posts
BS	3505	Specification for Unplasticised Polyvinyl Chloride (PVC-U) Pressure Pipes for Cold Potable Water
BS	3506	Specification for Unplasticised PVC Pipe for Industrial Uses
BS EN ISO	3506	Mechanical Properties of Corrosion-Resistant Stainless-Steel Fasteners
		Part 1: Bolts, Screws and Studs
		Part 2: Nuts
BS	3692	ISO Metric Precision Hexagon Bolts, Screws and Nuts. Specification
BS	3761	Specification for Solvent-based Paint Remover
BS	3837	Expanded Polystyrene Boards
		Part 1: Specification for Boards Manufactured from Expandable Beads
BS	3882	Specification for Topsoil
BS	3892	Pulverised-Fuel Ash
		Part 2: Specification for Pulverized-Fuel Ash to be Used as a Type I Addition
		Part 3: Specification for Pulverised-fuel Ash for Use in Cementitious Grouts

STD TYPE	STD NO	TITLE
BS	3936	Nursery Stock
		Part 1: Specification for Trees and Shrubs
BS	3969	Recommendations for Turf for General Purposes
BS	3998	Recommendations for Tree Work
BS	4027	Specification for Sulphate-resisting Portland Cement
BS	4043	Recommendations for Transplanting Root-balled Trees
BS	4190	ISO Metric Black Hexagon Bolts, Screws and Nuts. Specification
BS	4211	Specification for Ladders for Permanent Access to Chimneys, Other High Structures, Silos and Bins
BS	4213	Specification for Cold Water Storage and Combined Feed and Expansion Cisterns (Polyolefin Copolymer) up to 500L Capacity Used for Domestic Purposes
BS	4320	Specification for Metal Washers for General Engineering Purposes. Metric Series
BS	4395	Specification for High Strength Friction Grip Bolts and Associated Nuts and Washers for Structural Engineering Part 1: General Grade
BS	4449	Specification for Carbon Steel Bars for the Reinforcement of Concrete
BS	4482	Specification for Cold Reduced Steel Wire for the Reinforcement of Concrete
BS	4483	Specification for Steel Fabric for the Reinforcement of Concrete
BS	4504	Circular Flanges for Pipes, Valves and Fittings (PN Designated)
		Section 3.1: Specification for Steel Flanges
BS	4514	Unplasticised PVC Soil and Ventilating Pipes of 82.4 mm Minimum Mean Outside Diameter, and Fittings and Accessories of 82.4 mm and of Other Sizes. Specification
BS	4515	Specification for Welding of Steel Pipelines on Land and Offshore
		Part 1: Carbon and Carbon Manganese Steel Pipelines
		Part 2: Duplex Stainless Steel Pipelines
BS	4551	Methods of Testing Mortars, Screeds and Plasters
		Part 1: Physical Testing
		Part 1: Half-round Gutters and Pipes of Circular Cross-section
BS	4592	Industrial Type Metal Flooring, Walkways and Stair Treads
		Part 1: Specification for Open Bar Gratings
		Part 2: Specification for Expanded Metal Grating Panels
		Part 3: Specification for Cold Formed Planks
		Part 4: Specification for Glass Reinforced Plastics Open Bar Gratings
		Part 5: Solid Plates in Metal and Glass Reinforced Plastics (GRP). Specification
		Part 6: Glass Reinforced Plastics (GRP) Moulded Open Mesh Gratings and Protective Barriers. Specification
BS	4652	Specification for Zinc-Rich Priming Paint (Organic Media)
BS	4660	Thermoplastics Ancillary Fittings of Nominal Sizes 110 and 160 for Below Ground Gravity Drainage and Sewerage
BS	4729	Specification for Dimensions of Bricks of Special Shapes and Sizes
BS	4756	Specification for Ready Mixed Aluminium Priming Paints for Woodwork
BS	4787	Internal and External Wood Doorsets, Door Leaves and Frames
		Part 1: Specification for Dimensional Requirements
BS	4800	Schedule of Paint Colours for Building Purposes
BS	4841	Rigid Polyurethane Foam and Polyisocyanurate (PIR) for Building Applications
		Part 1: Specification for Laminated Board for General Purposes
		Part 2: Specification for Laminated Board for Use as a Wall and Ceiling Insulation
BS	4873	Aluminium Alloy Windows
BS	4880	Specification for Urinals
		Part 1: Stainless Steel Slab Urinals
BS	4962	Specification for Plastics Pipes and Fittings for Use as Sub Soil Field Drains
BS	4965	Specification for Decorative Laminated Plastics Sheet Veneered Boards and Panels
BS	4978	Specification for Softwood Grades for Structural Use
BS	4991	Specification for Propylene Copolymer Pressure Pipe
BS	5080	Structural Fixings in Concrete and Masonry
		Part 1: Method of Test for Tensile Loading
		Part 2: Method for Determination of Resistance to Loading in Shear
BS	5154	Specification for Copper Alloy Globe, Globe Stop and Check, Check and Gate Valves
BS	5158	Specification for Cast Iron Plug Valves
BS	5163	Specification for Predominantly Key-operated Cast Iron Gate Valves for Waterworks Purposes

STD TYPE	STD NO	TITLE
		Part 1: Code of Practice
		Part 2: Stem Caps for Use on Isolating Valves and Associated Water Control Apparatus. Specification
BS	5212	Cold Applied Joint Sealant Systems for Concrete Pavements
		Part 1: Specification for Joint Sealants
BS	5270	Bonding Agents for Use with Gypsum Plasters and Cement
		Part 1: Specification for Polyvinyl Acetate (PVAC) Emulsion Bonding Agents for Indoor Use with Gypsum Building Plasters
BS	5385	Wall and Floor Tiling
		Part 1: Design and Installation of Internal Ceramic Wall Tiling and Mosaics in Normal Conditions
		Part 2: Design and Installation of External Ceramic Wall Tiling and Mosaics (Including Terra Cotta and Faience Tiles)
		Part 3: Design and Installation of Ceramic Floor Tiles and Mosaics
		Part 4: Tiling and Mosaics in Specific Conditions
		Part 5: Design and Installation of Terrazzo Tile and Slab, Natural Stone and Composition Block Floorings
BS	5391	Specification for Acrylonitrile-butadiene-styrene (ABS) Pressure Pipe
		Part 1: Pipe for Industrial Uses
BS	5392	Specification for Acrylonitrile-butadiene-styrene (ABS) Fittings for Use with ABS Pressure Pipe
		Part 1: Fittings for Use with Pipe for Industrial Uses
BS	5412	Specification for Low-Resistance Single Taps and Combination Tap Assemblies (Nominal Size 1/2 and 3/4) Suitable for Operation at PN 10 Max. and a Minimum Flow Pressure of 0.01 MPa (0.1 Bar)
BS	5427	Code of Practice for the Use of Profiled Sheet for Roof and Wall Cladding on Buildings
		Part 1: Design
BS	5433	Specification for Underground Stopvalves for Water Services
BS	5520	Vitreous China Bowl Urinals (Rimless Type)
BS	5534	Code of Practice for Slating and Tiling
		Part 1: Design
BS	5606	Guide to Accuracy in Building
BS	5607	Code of Practice for the Safe Use of Explosives in the Construction Industry
BS	5627	Specification for Plastics Connectors for Use with Horizontal Outlet Vitreous China WC Pans
BS	5628	Code of Practice for Use of Masonry
		Part 3: Materials and Components, Design and Workmanship
BS	5642	Sills and Copings
		Part 1: Specification for Window Sills of Precast Concrete, Cast Stone, Clayware, Slate and Natural Stone
		Part 2: Specification for Copings of Precast Concrete, Cast Stone, Clayware, Slate and Natural Stone
BS	5709	Specification for Stiles, Bridle Gates and Kissing Gates
BS	5834	Surface Boxes, Guards and Underground Chambers for Gas and Waterworks Purposes
		Part 1: Specification for Guards, Including Foundation Units
		Part 2: Specification for Small Surface Boxes
		Part 3: Specification for Large Surface Boxes
		Part 4: Specification for Preformed Chambers
BS	5835	Recommendations for Testing of Aggregates
		Part 1: Compactibility Test for Graded Aggregates
BS	5837	Guide for Trees in Relation to Construction
BS	5911	Precast Concrete Pipes, Fittings and Ancillary Products
		Part 1: Specification for Unreinforced and Reinforced Concrete Pipes (Including Jacking Pipes) and Fittings With Flexible Joints (Complementary to BS EN 1916:2002)
		Part 3: Specification for Unreinforced and Reinforced Concrete Manholes and Soakaways (Complementary to BS EN 1917:2002)
		Part 5: Specification for Prestressed Non-pressure Pipes and Fittings With Flexible Joints
		Part 6: Specification for Road Gullies and Gully Cover Slabs
BS	5960	Specification for Minimum Information to be Included in Specification Sheets of Duplicators
BS	5975	Code of Practice for Falsework
BS	5977	Lintels
		Part 1: Method for Assessment of Load
BS EN ISO	5999	Polymeric Materials, Cellular Flexible. Polyurethane Foam for Load-bearing Applications

STD TYPE	STD NO	TITLE
		Excluding Carpet Underlay
BS	6031	Code of Practice for Earthworks
BS	6076	Specification for Tubular Polyethylene Film for Use as a Protective Sleeving for Buried Iron Pipes and Fittings
BS	6093	Code of Practice for Design of Joints and Jointing in Building Construction
BS	6150	Code of Practice for Painting of Buildings
BS	6164	Code of Practice for Safety in Tunnelling in the Construction Industry
BS	6180	Code of Practice for Protective Barriers in and about Buildings
BS	6187	Code of Practice for Demolition
BS	6213	Guide to the Selection of Constructional Sealants
BS	6262	Glazing for Buildings
BS	6398	Specification for Bitumen Damp-Proof Courses for Masonry
BS	6446	Specification for Manufacture of Glued Structural Components of Timber and Wood Based Panel Products
BS	6452	Beads for Internal Plastering and Dry Lining
		Part 1: Specification for Galvanised Steel Beads
BS	6510	Specification for Steel Windows, Sills, Window Boards and Doors
BS	6515	Specification for Polyethylene Damp-proof Courses for Masonry
BS	6576	Code of Practice for Installation of Chemical Damp-proof Courses
BS	6610	Specification for Pozzolanic Pulverised-Fuel Ash Cement
BS	6657	Guide to Prevention of Inadvertent Initiation of Electro-Explosive Devices by Radio-Frequency Radiation
PD	6682	Aggregates.
		Part 1: Aggregates for Concrete. Guidance on the Use of BS EN 12620
		Part 3: Aggregates for Mortar. Guidance on the Use of BS EN 13139
		Part 4: Lightweight Aggregates for Concrete, Mortar and Grout. Guidance on the Use of BS EN 13055-1
BS	6683	Guide to Installation and Use of Valves
BS	6700	Design, Installation, Testing and Maintenance of Services Supplying Water for Domestic Use Within Buildings and their Curtilages
BS	6744	Specification for Austenitic Stainless Steel Bars for the Reinforcement of Concrete
BS	6767	Transportable Accommodation Units
		Part 1: Recommendations for Design and Construction of the Basic Unit
BS	6920	Suitability of Non-Metallic Products for Use in Contact With Water Intended for Human Consumption With Regard to their Effect on the Quality of the Water Part 1: Specification
BS	6925	Specification for Mastic Asphalt for Building and Civil Engineering (Limestone Aggregate)
BS	7079	Preparation of Steel Substrates Before Application of Paints and Related Products. Introduction
BS	7361	Cathodic Protection
		Part 1: Code of Practice for Land and Marine Applications
BS	7371	Coatings on Metal Fasteners
		Part 12: Requirements for Imperial Fasteners
BS	7412	Plastics Windows Made From PVC-U Extruded Hollow Profiles
BS	7533	Guide for Structural Design of Pavements Constructed with Clay or Concrete Block Pavers
		Part 3: Pavements Constructed with Clay, Natural Stone or Concrete Pavers. Code of Practice for Laying Precast Concrete Paving Blocks and Clay Pavers for Flexible Pavements
		Part 4: Pavements Constructed with Clay, Natural Stone or Concrete Pavers. Code of Practice for the Construction of Pavements of Precast Concrete Flags or Natural Stone Slabs
BS EN ISO	7599	Anodising of Aluminium and its Alloys. General Specifications for Oxidation Coating of Aluminium
BS	7671	Requirements for Electrical Installations. IEE Wiring Regulations. Seventeenth Edition
BS	7775	Specification for Penstocks
BS	7786	Specification for Unsintered PTFE Tape for General Use
BS	7874	Method of Test for Microbiological Deterioration of Elastomeric Seals for Joints in Pipework and Pipelines
BS	7903	Guide to Selection and Use of Gully Tops and Manhole Covers for Installation Within the Highway
BS	7956	Specification for Primers for Woodwork
BS	7973	Spacers and Chairs for Steel Reinforcement and their Specification.
		Part 1: Product Performance Requirements
		Part 2: Fixing and Application of Spacers and Chairs and Tying of Reinforcement

STD TYPE	STD NO	TITLE
BS	8010	Pipelines
		Part 1: Pipelines on Land: General
		Section 2.1: Ductile Iron
		Section 2.4: Prestressed Concreted Pressure Pipelines
		Section 2.5: Glass Reinforced Thermosetting Plastics
		Section 2.7: Precast Concrete
		Section 2.8: Steel for Oil and Gas
BS	8102	Code of Practice for Protection of Structures Against Water from the Ground
BS ISO	8179	Ductile Iron Pipes - External Zinc based Coating
		Part 1: Metallic Zinc With Finishing Layer
BS	8203	Code of Practice for Installation of Sheet and Tile Flooring Resilient Floor Coverings
BS	8204	Screeds, Bases and In-Situ Floorings
		Part 2: Concrete Wearing Surfaces. Code of Practice
		Part 4: Cementitious Terrazzo Wearing Surfaces. Code of Practice
BS	8212	Code of Practice for Drylining and Partitioning Using Gypsum Plasterboard
BS	8215	Code of Practice for Design and Installation of Damp-Proof Courses in Masonry Construction
BS	8217	Code of Practice for Built-Up Felt Roofing
BS	8218	Code of Practice for Mastic Asphalt Roofing
BS	8313	Code of Practice for Accommodation of Building Services in Ducts
BS	8417	Preservation of Timber. Recommendations
BS	8481	Design, Preparation and Application of Internal Gypsum, Cement, Cement and Lime Plastering Systems. Specification
BS	8500	Concrete. (Complementary British Standard to BS EN 206-1)
		Part 1: Method of Specifying and Guidance for the Specifier
		Part 2: Specification for Constituent Materials and Concrete
BS	8666	Specification for Scheduling, Dimensioning, Bending and Cutting of Steel Reinforcement for Concrete
BS	9124	Specification for Steel and Aluminium Access Cover Systems With Over 1m Clear Opening
BS EN ISO	9934	Non-destructive Testing. Magnetic Particle Testing. General Principles
		Part 1: General Principles
BS EN ISO	9554	Fibre Ropes. General Specifications
BS	9990	Code of Practice for Non-automatic Fire-fighting Systems in Buildings
BS EN	10025	Hot Rolled Products of Structural Steels
		Part 1: General Technical Delivery Conditions
		Part 2: Technical Delivery Conditions for Non-alloy Structural Steels
BS EN	10056	Specification for Structural Steel Equal and Unequal Angles
		Part 1: Dimensions
BS EN	10080	Steel for the Reinforcement of Concrete. Weldable Ribbed Reinforcing Steel B500. Technical Delivery Conditions for Bars, Coils and Welded Fabric
BS EN	10088	Stainless Steels
		Part 3: Technical Delivery Conditions for Semi-finished Products, Bars Rods, Wire, Sections and Bright Products of Corrosion Resisting Steels for General Purposes
BS EN	10162	Cold Rolled Steel Sections. Technical Delivery Conditions. Dimensional and Cross-Sectional Tolerances
BS EN	10210	Hot Finished Structural Hollow Sections of Non-alloy and Fine Grain Structural Steels
		Part 2: Tolerances, Dimensions and Sectional Properties
BS EN	10224	Non-alloy Steel Tubes and Fittings for the Conveyance of Water and Other Aqueous Liquids. Technical Delivery Conditions
BS EN	10244	Steel Wire and Wire Products. Non-Ferrous Metallic Coatings on Steel Wire
		Part 2: Zinc or Zinc Alloy Coatings
BS EN	10255	Non-alloy Steel Tubes Suitable for Welding and Threading. Technical Delivery Conditions
BS EN	10296	Welded Circular Steel Tubes for Mechanical and General Engineering Purposes. Technical Delivery Conditions.
		Part 1: Non-Alloy and Alloy Steel Tubes
		Part 2: Stainless Steel
BS EN	10297	Seamless Circular Steel Tubes for Mechanical and General Engineering Purposes. Technical Delivery Conditions
		Part 1: Non-alloy and Alloy Steel Tubes
BS EN	10311	Joints for the Connection of Steel Tubes and Fittings for the Conveyance of Water and Other Aqueous Liquids

STD TYPE	STD NO	TITLE
BS EN ISO	11600	Building Construction. Jointing Products. Classifications and Requirements for Sealants
BS EN	12200	Plastics Rainwater Piping Systems for Above Ground External Use. Unplasticised Poly(Vinyl Chloride) (PVC-U) Part 1: Specifications for Pipes, Fittings and the System
BS EN	12201	Plastics Piping Systems for Water Supply. Polyethylene (PE) Part 1: General Part 2: Pipes Part 3: Fittings
BS EN	12288	Industrial Valves. Copper Alloy Gate Valves
BS EN	12326	Slate and Stone Products for Discontinuous Roofing and Cladding Part 1: Product Specification
BS EN	12334	Industrial Valves. Cast Iron Check Valves
BS EN	12588	Lead and Lead Alloys. Rolled Lead Sheet for Building Purposes
BS EN	12591	Bitumen and Bituminous Binders. Specifications for Paving Grade Bitumens
BS EN	12620	Aggregates for Concrete
BS EN	12828	Heating Systems in Buildings. Design for Water-based Heating Systems
BS EN	12831	Heating Systems in Buildings. Method for Calculation of the Design Heat Load
BS EN	12871	Wood-based Panels. Performance Specifications and Requirements for Load Bearing Boards for Use in Floors, Walls and Roofs
BS EN	12878	Pigments for the Colouring of Building Materials Based on Cement and/or Lime. Specifications and Methods of Test
BS EN ISO	12944	Paints and Varnishes. Corrosion Protection of Steel Structures by Protective Paint Systems
BS EN	13055	Lightweight Aggregates Part 1: Lightweight Aggregates for Concrete, Mortar and Grout
BS EN	13101	Steps for Underground Man Entry Chambers. Requirements, Marking, Testing and Evaluation of Conformity
BS EN	13108	Bituminous Mixtures. Material Specifications Part 1: Asphalt Concrete Part 4: Hot Rolled Asphalt Part 7: Type Testing
BS EN	13121	GRP Tanks and Vessels for Use Above Ground Part 3: Design and Workmanship
BS EN	13139	Aggregates for Mortar
BS EN	13164	Thermal Insulation Products for Buildings. Factory Made Products of Extruded Polystyrene Foam (XPS). Specification
BS EN	13244	Plastics Piping Systems for Buried and Above-Ground Pressure Systems for Water for General Purposes, Drainage and Sewerage. Polyethylene (PE) Part 1: General
BS EN	13279	Gypsum. Binders and Gypsum Plasters Part 1: Definitions and Requirements
BS EN	13310	Kitchen Sinks. Functional Requirements and Test Methods
BS EN	13369	Common Rules for Precast Concrete Products
BS EN	13397	Industrial Valves. Diaphragm Valves Made of Metallic Materials
BS EN	13476	Plastics Piping Systems for Non-pressure Underground Drainage and Sewerage. Structured-wall Piping Systems of Unplasticized Poly(Vinyl Chloride) (PVC-U), Polypropylene (PP) and Polyethylene (PE) Part 1: General Requirements and Performance Characteristics Part 2: Specification for Pipes and Fittings Part 3: Specifications for Pipes and Fittings With Smooth Internal and Profiled External Surface and the System, Type B
BS EN	13566	Plastics Piping Systems for Renovation of Underground Non-Pressure Drainage and Sewerage Networks Part 1 General Part 2 Lining with Continuous Pipes Part 3 Lining with Close Fit Pipes Part 4: Lining with Cured-in-Place Pipes
BS EN	13598	Plastics Piping Systems for Non-pressure Underground Drainage and Sewerage. Unplasticised Poly(Vinyl Chloride) (PVC-U), Polypropylene (PP) and Polyethylene (PE) Part 1: Specifications for Ancillary Fittings Including Shallow Inspection Chambers Part 2: Specifications for Manholes and Inspection Chambers in Traffic Areas and Deep Underground Installations
BS EN	13658	Metal Lath and Beas. Definitions, Requirements and Test Methods

STD TYPE	STD NO	TITLE
		Part 1: Internal Plastering
		Part 2: External Rendering
BS EN	13670	Execution of Concrete Structures
BS EN	13706	Reinforced Plastics Composites. Specifications for Pultruded Profiles
		Part 2: Method of Test and General Requirements
		Part 3: Specific Requirements
BS EN	13707	Flexible Sheets for Waterproofing. Reinforced Bitumen Sheets for Roof Waterproofing. Definitions and Characteristics
BS EN	13748	Terrazzo Tiles
		Part 1: Terrazzo Tiles for Internal Use
		Part 2: Terrazzo Tiles for External Use
BS EN	13877	Concrete Pavements
		Part 3: Specifications for Dowels to be Used in Concrete Pavements
BS EN	13914	Design, Preparation and Application of External Rendering and Internal Plastering
		Part 1: External Rendering
		Part 2: Design Considerations and Essential Principles for Internal Plastering
BS EN	13923	Filament-wound FRP Pressure Vessels. Materials, Design, Manufacturing and Testing
BS EN	14020	Reinforcements. Specification for Textile Glass Rovings
		Part 2: Methods of Test and General Requirements
		Part 3: Specific Requirements
BS EN ISO	14122	Safety of Machinery. Permanent Means of Access to Machinery
		Part 2: Working Platforms and Walkways
		Part 3: Stairways, Stepladders and Guard-rails
		Part 4: Fixed Ladders
BS EN	14161	Petroleum and Natural Gas Industries. Pipeline Transportation Systems
BS EN ISO	14171	Welding Consumables. Solid Wire Electrodes, Tubular Cored Electrodes and Electrode/Flux Combinations for Submerged Arc Welding of Non Alloy and Fine Grain Steels. Classification
BS EN	14188	Joints Fillers and Sealants
		Part 1: Specifications for Hot Applied Sealants
BS EN	14336	Heating Systems in Buildings. Installation and Commissioning of Water Based Heating Systems
BS EN	14339	Underground Fire Hydrants
BS EN ISO	14341	Welding Consumables. Wire Electrodes and Deposits for Gas Shielded Metal Arc Welding of Non Alloy and Fine Grain Steels. Classification
BS EN ISO	14343	Welding Consumables. Wire Electrodes, Strip Electrodes, Wires and Rods for Arc Welding of Stainless and Heat Resisting Steels. Classification
BS EN	14364	Plastics Piping Systems for Drainage and Sewerage With of Without Pressure. Glass-reinforced Thermosetting Plastics (GRP) Based on Unsaturated Polyester Resin (UP). Specifications for Pipes, Fittings and Joints
BS EN	14396	Fixed Ladders for Manholes
BS EN	14399	High-strength Structural Bolting Assemblies for Preloading
BS EN	14411	Ceramic Tiles. Definitions, Classification, Characteristics and Marking
BS ISO	14654	Epoxy-Coated Steel for the Reinforcement of Concrete
BS EN	14680	Adhesives for Non-pressure Thermoplastic Piping Systems. Specifications
BS EN	14686	Fibre Ropes of Polyester/Polyolefin Dual Fibres
BS EN ISO	14713	Protection Against Corrosion of Iron and Steel in Structures. Zinc and Aluminum Coatings. Guidelines
		Part 1: Zinc Coatings. Guidelines and Recommendations for the Protection Against Corrosion of Iron and Steel in Structures. General Principles of Design and Corrosion Resistance
BS EN	14782	Self-supporting Metal Sheet for Roofing, External Cladding and Internal Lining. Product Specification and Requirements
BS EN	14814	Adhesives for Thermoplastic Piping Systems for Fluids Under Pressure. Specifications
BS EN	14844	Precast Concrete Products. Box Culverts
PD CEN/TR	15123	Design, Preparation and Application of Internal Polymer Plastering Systems
BS EN	15167	Ground Granulated Blast Furnace Slag for Use in Concrete, Mortar and Grout
		Part 1: Definitions, specifications and conformity criteria
BS EN	15282	Vitreous and Porcelain Enamels. Design of Bolted Steel Tanks for the Storage or Treatment of Water or Municipal or Industrial Effluents and Sludges
BS ISO	15686	Buildings and Constructed Assets. Service Life Planning
		Service Life Prediction Procedures

STD TYPE	STD NO	TITLE
BS ISO	16132	Ductile Iron Pipes and Fittings. Seal Coats for Cement Mortar Linings
BS EN ISO	18273	Welding Consumables. Wire Electrodes, Wires and Rods for Welding of Aluminium and Aluminium Alloys. Classification
BS EN	61386	Conduit Systems for Cable Management. Particular Requirements
		Part 24: Conduit Systems Buried Underground
BS	594987	Asphalt for Roads and Other Paved Areas. Specification for Transport, Laying, Compaction and Type Testing Protocols

APPENDIX III
LIST OF WATER INDUSTRY SPECIFICATIONS/INFORMATION AND GUIDANCE NOTES TO WHICH REFERENCE IS MADE IN THE DOCUMENT

All documents are available from Water UK at www.water.org.uk/home/member-services/wis-and-ign.

WIS/IGN	NUMBER	TITLE
IGN	4-01-03	Field Pressure Testing of Pressure Pipes and Fittings
WIS	4-02-01	Operational Requirements: *In situ* Resin Lining of Water Mains
IGN	4-02-02	Code of Practice: *In situ* Resin Lining of Water Mains
IGN	4-08-01A	Bedding and Sidefill Materials for Buried Pipelines
WIS	4-08-02A	Specification for Bedding and Sidefill Materials for Buried Pipelines
IGN	4-11-01	Vitrified Clay Pipes and Fittings
WIS	4-12-04	Specification for Glassfibre Reinforced Cement (GRC) Sewer Linings
WIS	4-12-05	Specification for Precast Gunite Sewer Linings
WIS	4-12-06	Specification for Precast and In-situ Ferrocement
WIS	4-13-01	Specification for Blastfurnace Slag Cement for In-situ Lining of Water Mains
IGN	4-21-01	Ductile Iron Pipes and Fittings
WIS	4-21-02	Mechanical Couplings and Repair Clamps for Iron Pipes for the Conveyance of Cold Potable Water (Underground Use) for the Size Range 40 to 1600 mm/1.5 to 48" Inclusive
WIS	4-22-02	Specification for Ferrules (Tapping Tees) and Ferrule Straps for Underground Use
WIS	4-23-04	Specification for Underground Stopvalves, Including Spherical Valves, for Potable Water Services for Nominal Sizes up to and Including 63 and Nominal Pressures of 10 Bar Minimum and Made Principally of Metal or Thermoplastics
WIS	4-24-01	Specification for Mechanical Fittings and Joints Including Flanges for PE Pipes for the Conveyance of Cold Potable Water for the Size Range 90-1000 Made of Metal or Plastics or a Combination of Both
WIS	4-25-01	Specification for the Use of Steel Tanks in the Water Industry
WIS	4-31-08	Specification for Molecular Oriented Polyvinyl Chloride (MOPVC) Pressure Pipes for Underground Use
WIS	4-32-05	Specification for Polyethylene (PE) Pipes for Sewer Lining (Non-pressure Applications)
WIS	4-32-08	Specification for Site Fusion Jointing of Polyethylene Pressure Pipeline Systems Using PE80 and PE100 Materials
WIS	4-32-11	Specification for Thermoplastic End Load Resistant Mechanical Fittings for Polyethylene Pipes of Nominal Size ≤63
WIS	4-32-16	Butt Fusion Jointing Machines
IGN	4-32-18	The Choice of Pressure Ratings for Polyethylene Pipe Systems for Water Supply and Sewerage Duties
WIS	4-34-02	Specification for Glassfibre Reinforced Plastics (GRP) Sewer Linings
WIS	4-34-04	Specification for Renovation of Gravity Sewers by Lining with Cured-in-place Pipes
WIS	4-34-05	Specification for Polyester Resin Concrete (PRC) Sewer Linings
WIS	4-34-06	Specification for Localised Sewer Repairs Using Cured-in-place System With or Without Re-rounding
WIS	4-35-01	Specification for Thermoplastics Structured Wall Pipes - Supplementary Test Requirements
WIS	4-37-01A	Specification for Boundary Boxes for the Metering and Control of Domestic and Small Industrial Water Services (Amendment)
WIS	4-41-01	Specification for Flexible Couplings for Gravity Sewerage and Drainage Pipes
IGN	4-50-03	Operating Guidelines for the Use of Site-applied, Factory Applied, and Reinforced Factory Applied Polyethylene Sleeving on Ductile Iron Pipeline Systems

WIS/IGN		NUMBER TITLE
IGN	4-51-01	External Zinc Coating of Ductile Iron Pipe
WIS	4-52-01A	Specification for Polymeric Anti-corrosion (Barrier) Coatings (Amendment)
IGN	4-52-02	The Use of Polymeric Anti-corrosion (Barrier) Coatings
WIS	4-52-03A	Anti-corrosion Coatings on Threaded Fasteners (Amendment)

APPENDIX IV

LIST OF CONSTRUCTION INDUSTRY RESEARCH AND INFORMATION ASSOCIATION (CIRIA) PUBLICATIONS TO WHICH REFERENCE IS MADE IN THE DOCUMENT

NUMBER	TITLE
C511	Controlled Permeability Formwork
C512	Environmental Handbook for Building and Civil Engineering Projects. Part 1: Design and Specification
C528	Environmental Handbook for Building and Civil Engineering Projects. Part 2: Construction
C648	Control of Water Pollution from Linear Construction Projects.Technical Guidance
C649	Control of Water Pollution from Linear Construction Projects. Site Guide.
C650	Environmental Good Practice on Site (2nd Edition)
C660	Early-age Thermal Crack Control in Concrete
C698	Site Handbook for the Construction of SUDS
R44	Medical Code of Practice for Work in Compressed Air, (3rd Edition) 1982
R59	Building Sands: Availability, Usage and Compliance with Specification Requirements
R97	Trenching Practice (2nd Edition)
R116	Design of Reinforced Grass Waterways
R128	Guide to the Design of Thrust Blocks for Buried Pressure Pipelines
R136	Formwork Striking Times – Criteria, Prediction and Methods of Assessment
SP116	Environmental Impact of Building and Construction Materials, Volume A. Summary
TN75	Load Tests on Fixings in Concrete
TN95	Proprietary Trench Support Systems (3rd Edition)
TN104	Precast Concrete Tunnel Linings – A Review of Current Test Procedures
TN128	Civil Engineering Sealants in Wet Conditions – Review of Performance and Interim Guidance on Use
TN137	Selection and Use of Fixings in Concrete and Masonry - an Update to CIRIA Guide 4
TN144	Performance of Sealant-concrete Joints in Wet Conditions - Results of a Laboratory Testing Programme.Volume 1: Main Results and Discussion

APPENDIX V
LIST OF STATUTORY REFERENCES IN THE DOCUMENT

Building Regulations 2000

Construction (General Provisions) Regulations 1961

Construction (Health, Safety and Welfare) Regulations 1996

Construction (Health, Safety and Welfare) Regulations (Northern Ireland) 1996

Construction (Metrication) Regulations 1984

Control of Asbestos Regulations 2006

Control of Asbestos Regulations (Northern Ireland) 2007

Control of Explosives Regulations 1991

Control of Pollution (Oil Storage) (England) Regulations 2001

Control of Pollution (Oil Storage) Regulations (Northern Ireland) 2010

Council Directive on the Approximation of Laws, Regulations and Administrative Provisions of the Member States Relating to Construction Products (89/106/EEC)

Council Directive on the Procurement Procedures of Entities in the Water, Energy, Transport and Telecommunications Sectors (93/38/EEC)

Dangerous Substances and Explosive Atmospheres Regulations 2002

Dangerous Substances and Explosive Atmospheres Regulations (Northern Ireland) 2003

Environmental Protection Act 1990 (amended by Environment Act 1995)

Explosives Acts 1875 and 1923

Financing Act 1996 (Landfill Tax), amended by the Finance Act 1999 (and subsequent updates)

Hedgerows Regulations 1997

Land Drainage Act 1994

Manual Handling Operations Regulations 1992

New Roads and Street Works Act 1991

Petroleum (Consolidation) Act 1928

Petroleum (Consolidation) Act (Northern Ireland) 1929

Petroleum Spirit (Motor Vehicles etc.) Regulations 1929

Planning (Control of Major-accident Hazards) Regulations 2009

Planning (Control of Major-accident Hazards) Regulations (Northern Ireland) 2009

Planning (Control of Major-accident Hazards) (Scotland) Regulations 2009

Planning (Hazardous Substances) Regulations 1992

Planning (Hazardous Substances) (Amendment) (England) Regulations 2009

Planning (Hazardous Substances) (Amendment) Regulations (Northern Ireland) 2005

Salmon and Fresh Water Fisheries Act 1975

Scottish Water Byelaws (2004)

Site Waste Management Plan Regulations 2008

Street Works (Northern Ireland) Order 1995 and amendments

Town and Country Planning (Hazardous Substances) (Scotland) Regulations 1993

Traffic Management Act 2004 (and Orders for England and Wales)

Traffic Signs Regulations and General Directions 2002

Utilities Directive on the Procurement Procedures of Entities Operating in the Water, Energy, Transport and Postal Services Sectors (2004/17/EC)

Water Resources Act 1991

Water Supply (Water Fittings) Regulations (Northern Ireland) 2009

Water Supply (Water Fittings) Regulations (1999)

Water Supply (Water Quality) (Scotland) Regulations 2001

Water Supply (Water Quality) Regulations 2000

Water Supply (Water Quality) Regulations 2010

Water Supply (Water Quality) Regulations (Northern Ireland) 2002 and amendments 2003

Water Supply (Water Quality) Regulations (Wales) 2001

Wildlife and Countryside Act 1981 and amendments

Work in Compressed Air Regulations 1996

Work in Compressed Air Regulations (Northern Ireland) 2004

APPENDIX VI

LIST OF OTHER PUBLICATIONS TO WHICH REFERENCE IS MADE IN THE DOCUMENT

Alkali-Silica Reaction, Digest 330, 1999. Building Research Establishment (BRE), Bucknalls Lane, Garston, Watford, WD2 7JR

ASTM D3575.-00e1 American Society for Testing Materials, 1916 Race Street, Philadelphia, PA 19103, USA

Avoiding Danger from Underground Services, HSG 47, HSE, PO Box 1999, Sudbury, Suffolk, CO10 6FS

Bricklaying in Winter Conditions. Brick Development Association

Chamber Tops and Gully Tops for Road Drainage and Services: Installation and Maintenance, 2009. HA 104/09. Highways Agency

Concrete in Aggressive Ground, Special Digest No. 1, 2005. Building Research Establishment, Bucknalls Lane, Garston, Watford, WD2 7JR

Damage Control Procedures for Pipeline Construction Involving Pipe Splitting. UKWIR/British Gas (Currently Transco). UKWIR Ltd, 1 Queen Anne's Gate, London, SW1H 9BT

Every Drop Counts – Keeping Water Clean. Crop Protection Association, 4 Lincoln Court, Lincoln Road, Peterborough, PE1 2RP (available at - http://www.voluntaryinitiative.org.uk/_Attachments/Resources/978_S4.pdf)

Foamed Concrete - A Specification for Use in Reinstatement of Openings in Highways. British Cement Association, Century House, Telford Avenue, Crowthorne, Berkshire, RG11 6YS

Guidance for the Design of Steel Fibre Reinforced Concrete. Technical Report 63. The Concrete Society, Riverside House, 4 Meadows Business Park, Station Approach, Blackwater, Camberley, Surrey, GU17 9AB

Guidance for the Selection of Water Supply Pipes to be used in Brownfield Sites. UKWIR Ltd, 1 Queen Anne's Gate, London, SW1H 9BT

Guide to Best Practice for the Installation of Pipejacks and Microtunnels. Pipe Jacking Association, Hamilton House, 1 Temple Avenue, London, EC4Y 0HA

Guide to Site Practice. Box Culvert Association, 60 Charles Street, Leicester LE1 1FB

Guidelines for the Planning, Installation and Maintenance of Utility Services in Proximity to Trees, NJUG Publication No.10, 1995. National Joint Utilities Group, 30 Millbank, London, SW1P 4RD

Hardcore, Digest 276. Building Research Establishment (BRE), Bucknalls Lane, Garston, Watford, WD2 7JR

Horizontal Directional Drilling Good Practice Guidelines (3rd Edition). North American Society for Trenchless Technology

In-situ Cement Mortar Lining - Operational Guidelines and Code of Practice. WRc plc, Frankland Road, Blagrove, Swindon, Wiltshire, SN5 8YF

Leaving Decommissioned Asbestos Pipes in Excavations. Environment Agency (available at http://environment-agency.co.uk)

Model Consultative Procedure for Pipeline Construction Involving Deep Excavation, 1993. British Gas/WSA/WCA, Ashby Road, Loughborough, LE11 3GR

Model Contract Document for Sewer Condition Inspection, WSA and FWR, 1994. WRc plc, Frankland Road, Blagrove, Swindon, Wiltshire, SN5 8YF

National Structural Steelwork Specification for Building Construction, 2009. British Constructional Steelwork Association

Occupational Exposure Limits, Guidance Note EH40, 2002 plus supplement 2003. HSE, PO Box 1999, Sudbury, Suffolk, CO10 6FS

Pollution Prevention Guidance Notes. Environment Agency (available at www.environment-agency.gov.uk)

Pollution Prevention Guidelines PPG2 and PPG 26)

Polyethylene Pipe Systems Manual. WRc plc/ BPF (available at www.wrcplc.co.uk/pe)

Principles of Water Supply Hygiene: Technical Guidance Notes, 2010. Water UK/Royal Society for Public Health (available at www.water.org.uk)

Procedures for Testing the Capability of Cured In Place Pipe (CIPP) Lining Systems to Resist Infiltration, Report CP308. WRc plc, Frankland Road, Blagrove, Swindon, Wiltshire, SN5 8YF

Requirements for Electrical Installations, BS 7671: 2001. Institution of Electrical Engineers, Savoy Place, London, WC2R 0BL

Safety at Street Works and Road Works, 2002. The Stationery Office Ltd, PO Box 276, London, SW8 5DT

Specification for Portable Traffic Signal Control Equipment for Use at Roadworks TR 2502

Specification for the Reinstatement of Openings in Highways. Department for Transport (available at www.dft.gov.uk)

Specification for the Reinstatement of Openings in Highways (2nd Edition)

Specification for the Reinstatement of Openings in Roads.The Scottish Government (available at www.scotland.gov.uk)

Specification for Tunnelling. British Tunnelling Society/ Institution of Civil Engineers

Sewerage Rehabilitation Manual (5th Edition), 2001. WRc plc (available at http://srm.wrcplc.co.uk)

Sliplining and Replacement Moling of Water Mains – Operational Guidelines. WRc plc, Frankland Road, Blagrove, Swindon, Wiltshire, SN5 8YF

Specification for Highway Works. Highways Agency, 123 Buckingham Palace Road, London, SW1W 9HA

Specification for Piling and Embedded Retaining Walls. Institution of Civil Engineers, 1-7 Great George Street, London, SW1P 3AA

Specification for Tunnelling. British Tunnelling Society and Institution of Civil Engineers, 1-7 Great George Street, London, SW1P 3AA

The Management of Asbestos in Non-domestic Premises, Regulation 4 of the Control of Asbestos Regulations 2006. Health and Safety Executive (available at www.hse.gov.uk)

Traffic Signs Manual. Department of Transport, The Stationery Office Ltd, PO Box 276, London, SW8 5DT

Traffic Signs Regulations and General Directions, 2002. The Stationery Office Ltd, PO Box 276, London, SW8 5DT

Typical Specifications for Trenchless Technology Systems, UM 1464, 1995

Water Fittings and Materials Directory (www.wras.co.uk/directory)

Water Mains Rehabilitation Manual, 1989. WRc plc, Frankland Road, Blagrove, Swindon, Wiltshire, SN5 8YF

Winter Concreting and Technical Guidance Notes, 1985. Concrete Information Ltd, Century House, Telford Avenue, Crowthorne, Berkshire, RG11 6Y

Work with Materials Containing Asbestos. Control of Asbestos Regulations 2006. Health and Safety Executive (available at www.hse.gov.uk)

INDEX TO SPECIFICATION

Clause